VHDL: A Starter's Guide

Second Edition

Sudhakar Yalamanchili
Georgia Institute of Technology

PEARSON

Prentice
Hall

Upper Saddle River, NJ 07458

Library of Congress Cataloging-in-Publication Data on File

Vice President and Editorial Director, ECS: *Marcia J. Horton*
Vice President and Director of Production and Manufacturing, ESM: *David W. Riccardi*
Executive Managing Editor: *Vince O'Brien*
Managing Editor: *David A. George*
Production Editor: *Scott Disanno*
Director of Creative Services: *Paul Belfanti*
Art Director: *Jayne Conte*
Cover Design: *Bruce Kenselaar*
Art Editor: *Greg Dulles*
Manufacturing Buyer: *Lisa McDowell*
Senior Marketing Manager: *Holly Stark*

© 2005, 1998 Pearson Education, Inc.
Pearson Prentice Hall
Pearson Education, Inc.
Upper Saddle River, NJ 07458

Pearson Prentice Hall™ is a trademark of Pearson Education, Inc.

Active-HDL is a trademark of Aldec, Inc., 2260 Corporate Circle Henderson, NV 89074

The author and publisher of this book have used their best efforts in preparing this book. These efforts include the development, research, and testing of the theories and programs to determine their effectiveness. The author and publisher make no warranty of any kind, expressed or implied, with regard to these programs or the documentation contained in this book. The author and publisher shall not be liable in any event for incidental or consequential damages in connection with, or arising out of, the furnishing, performance, or use of these programs.

Printed in the United States of America

10 9 8 7 6 5 4 3 2 1

ISBN 0-13-145735-7

Pearson Education Ltd., *London*
Pearson Education Australia Pty. Ltd., *Sydney*
Pearson Education Singapore, Pte. Ltd.
Pearson Education North Asia Ltd., *Hong Kong*
Pearson Education Canada, Inc., *Toronto*
Pearson Educación de Mexico, S.A. de C.V.
Pearson Education—Japan, *Tokyo*
Pearson Education Malaysia, Pte. Ltd.
Pearson Education, Inc., *Upper Saddle River, New Jersey*

Dedicated to the memory of

John Uyemura

Mentor, Colleague, and Inspiration

Contents

Preface

Hardware description languages (HDLs) are a staple component of modern digital design flows, and thus it is imperative that we provide opportunities for Electrical Engineering, Computer Engineering, and Computer Science students to learn the basic concepts underlying HDLs and to become proficient in their application. While many hardware description languages have been proposed and are in active use for a variety of specific digital design tasks, at the time of this writing the VHDL and Verilog languages remain the dominant languages in industry. The acronym VHDL stands for **VHSIC Hardware Description Language** (more on the history in the introduction), and the language enjoys support from every major vendor of computer-aided design tools.

There are many excellent texts on the description of the VHDL language and its use for the purpose of building accurate models of complex digital systems. These texts have been largely complete treatments of the language and generally have been written for practicing engineers and, more recently, also for use in graduate courses. This text has a different goal. It is not intended to be a comprehensive VHDL language reference. Rather, it is intended to provide an introduction to the *basic* language concepts and a framework for *thinking about* the structure and operation of VHDL programs. Programming idioms from conventional programming languages by themselves are insufficient for productively learning to apply hardware description languages such as VHDL. Through simulation and laboratory exercises, students can very quickly come up to speed in building useful, nontrivial models of digital systems. As their experience grows, so will their need for more comprehensive information and modeling techniques, which can be found in a variety of existing texts and the standard language reference manual.

The VHDL language can be used with several goals in mind. It may be used for the synthesis of digital circuits, verification and validation of digital designs, test vector generation for testing circuits, or simulation of digital systems. The use of the language for each of these purposes will emphasize certain of its aspects. This text is intended to support a first look at the language, and, as such, I have decided to adopt the application to the simulation of digital systems in introducing the basic language concepts. Digital logic synthesis is an alternative, widely used application of the language. A good introduction to the use of VHDL for logic synthesis can be found in several texts (for example see [3, 17]).

Intended Audience

The style of the book is motivated by the need for a companion text for sophomore- and junior-level texts used in courses in digital logic and computer architecture. The ability to construct simulation models of the building blocks studied in these courses is an invaluable teaching aid. VHDL is a complex language that could easily be worthy of a course in its own right. However, curricula are usually strapped for credit hours, and devoting a course to teach VHDL would mean eliminating existing material. The style of this text is intended to permit integration of the basic concepts underlying VHDL into existing courses without necessitating additional credit hours or courses for instruction.

Students learn the most about digital systems if they have to build them, in this case using simulation models. I have found it valuable in my own courses to provide concurrent laboratory exercises that reinforce foundational material taught in the classroom. The scope and complexity of hardware laboratory exercises are limited by the available time in a semester or quarter-long course. The VHDL language provides an opportunity for students to experiment with larger designs than would be feasible in a hardware laboratory, using a development environment that is used throughout the industrial community. In the case of VHDL, I would like an approach that would enable students to adjust quickly to the basic language concepts such that they could construct models of basic logic and computer architecture components productively in sophomore- and junior-level courses. The full power of the language is not necessary at this point, nor should it be. As they progress to more advanced courses and their needs grow, students would be able to use productively the more advanced language features with their corresponding texts as references.

This book attempts to develop an intuition and a structured way of thinking about VHDL models without necessarily spending a great deal of time on advanced language features. Students should be able to learn enough quickly through exercises and association with classroom concepts to be able to construct useful models. This book strives to fill this need. In the past, the material in this text has been used as an adjunct for a two-quarter sequence on computer architecture at Georgia Tech. Sophomore students start with no background in VHDL but with a good background in high-level programming languages such as C or Pascal. By the end of the second quarter, they will have modified a simple, single-cycle datapath into a functional model of a pipelined RISC processor with hazard detection, data forwarding, and branch prediction. Early in the second quarter, VHDL makes the transition from a new language to a tool for studying computer architecture. My goal has been to enable early integration of VHDL into the curriculum, in a manner that strengthens the learning of the concepts while concurrently providing training in the use of VHDL simulation tools.

Style of the Book

In order to fill the need for a companion text for computer architecture courses and an early introduction to the basic language concepts, the book must satisfy several criteria. First, it must relate VHDL concepts to those already familiar to the student. Students learn best when they can relate new concepts to ones with which they are already familiar. In this case, we rely on concepts from the operation of digital circuits. Language features are motivated by the need to describe specific aspects of the operation of digital circuits—for example, events, propagation delays, and concurrency.

Second, each language feature must be accompanied by examples. Simulation exercises address one or more VHDL modeling concepts. In keeping with the idea of a companion book, a tutorial for a popular VHDL simulator is provided in Appendix A. Most modern VHDL simulators share a similar "look and feel," and many vendors provide inexpensive student editions that retain much of the functionality of the simulators while generally restricting the size of the problems, which is typically not an impediment to the goals of this text. Finally, the text must be *prescriptive*. Chapter 3, Chapter 4, and Chapter 5 each provide a prescription for writing classes of VHDL models. This is not intended to produce the most efficient models, but to rapidly bring the student to a point where he or she can construct useful simulation models for instructional purposes. By enabling a look at the detailed operation of digital systems, VHDL reinforces the foundational concepts taught in the classroom. At this point, students begin to think about alternative, more creative, and often more efficient approaches to constructing the models.

The approach taken is a bit unusual in that I do not begin with a discussion and presentation of language syntax and constructs (i.e., identifiers, operators, and so forth). In fact, the book presents core constructs via examples, and the syntax is not presented until late in the text, with the notion that this chapter will be used more as a reference. The premise is that readers who have had experience with programming in high-level languages simply need an accurate syntactic reference to these language constructs. The road to building useful models is built on an understanding of how we can describe those aspects of hardware systems that require constructs that are not typically found in traditional programming language definitions, such as signals and the concept of time. This is where the bulk of this book is focused. As a result, a syntactic reference to the core programming language features has been reduced to a single chapter. The goal is to focus on the concepts underlying the VHDL language and its use in simulation. If we can capture the novel features of the language in a manner that appeals to the reader's intuition and is based on thinking in hardware or systems terms, then a reference to the syntax of core constructs is sufficient to get students started in building useful models. My hope is that this text can apply this approach successfully and fulfill the goal of getting students at the sophomore level excited about the use of such languages in general and the evolving design methodologies in particular. They are then capable of more rapidly expanding their understanding to the full scope of the language.

Organization of the Text

As a result of this view, the text starts with concepts from the operation of digital circuits. This text assumes that the reader is comfortable with introductory digital logic and programming in a block-structured, high-level language such as Pascal or C. The subsequent chapters introduce various VHDL constructs as representations of the operational and physical attributes of digital systems. An association is made between the physical phenomena found in digital circuits and their representation within the VHDL language. Though VHDL is often criticized as a bulky and complex language in its entirety, these associations are intuitive and therefore easy to pick up. Once the basic concepts are understood, a good syntactic reference to the language is sufficient for the students to be able to quickly build models of digital circuits, including higher-level objects such as register files, ALUs, and simple datapaths.

Chapter 3 is perhaps the most important chapter in the book. The basic attributes of digital systems are associated with language features, and an overall structure of a VHDL model is developed. Chapter 4 and Chapter 5 build on this basic structure to introduce the concepts of processes, hierarchy, and abstraction. Each of these chapters provides simulation exercises that can be used by the student to reinforce the concepts.

Completion of the simulation exercises enables the student to proceed to more complex language features and to productive use of any of the existing comprehensive VHDL language texts. Chapter 6 and Chapter 7 present language features as additional functionality that serve a specific purpose—for example, input–output and procedures. By this time, students are able to write and simulate simple models. Thus, students can concentrate on each of these new features and their utilities. A syntactic reference to the common language features is provided in Chapter 9.

Several appendices have been added to support the material in the text. For example, Appendix A provides a tutorial for a popular VHDL simulator. Appendix B provides a handy reference to common VHDL packages. Finally, Appendix C provides a detailed template for a VHDL model illustrating the relative ordering of program constructs. This can serve as a handy reference toward the end of the student's experience.

By providing a companion text to enable students to learn VHDL "as they go," I hope that the text will enable more curricula to introduce VHDL early and productively while serving as a teaching aid for classroom material. Students are thus prepared for advanced courses dealing with state-of-the-art techniques, such as rapid prototyping of digital systems, high-level synthesis, and advanced modeling.

Caveats

This book is based on the 1993 VHDL standard, a revision of the 1987 standard. This text is based on the idea that at the sophomore and junior levels it is best to limit coverage to those features of the language that can be used to build useful simulation models that provide insights into the behavior of digital systems, particularly at the architectural level. Students can subsequently move on to a broader coverage of the language in many excellent textbooks, some of which are listed in the bibliography. As a result, the reader should be aware that this text does not provide complete coverage of the language features. Furthermore, the text introduces the IEEE 1164 standard data types early on and uses them in all of the examples rather than staying with the relatively simpler **bit** and **bit_vector** types. I hope that this will expose students to standard practice and not compromise the goals of focusing on core language features and types.

Acknowledgments

As with most textbook endeavors, this text grew out of a perceived need in the classroom, the encouragement of colleagues, the participation and feedback from students, and the never-ending accommodations of my family. It is a pleasure to acknowledge the contributions of the many people who contributed to its generation.

I am particularly grateful to Tom Robbins of Prentice Hall for his seemingly endless patience, his commitment to education, and his proactive demeanor in helping me complete these projects. Tom always gets to the heart of the matter in a way that is indicative of his insight into the educational needs of our students.

I am especially grateful to those colleagues who were generous with their time in reading early drafts of the manuscript and who were forthcoming with their comments and suggestions. This second edition is a product of those who contributed to the first edition and those who carried on the selfless tradition of critical feedback. Todd Carpenter, Vijay Madisetti, Linda Wills, Jay Schlag, and Abhijit Chatterjee commented and utilized early drafts in their classes. I am particularly grateful to numerous sophomore and junior students for their frank and forthright appraisal of the material and for the opportunity to gain their feedback. Mr. Subramanian Ramaswamy was very helpful in proofing drafts, in the development of the solutions manual, and in the critical assessment and formatting of supporting material for the text.

This book is dedicated to the late Professor John Uyemura. John was inspirational in getting me to believe that I could create a text that would fulfill a need in the classroom. He provided the initial impetus for the book as well as support and input throughout the project. His experience with how students learn was invaluable in shaping the approach employed in the text. He was a mentor throughout and inspirational in boosting my confidence in the project. While his presence will be deeply missed, his impact and influence will be timeless.

This book is a shared project. My colleagues, students, and family share any successes from the completion of this text, while any omissions and errors remain solely with the author.

SUDHAKAR YALAMANCHILI
Atlanta, Georgia

CHAPTER 1

Introduction

1.1 What Is VHDL?

The acronym VHDL stands for the **V**HSIC **H**ardware **D**escription **L**anguage. The acronym VHSIC, in turn, refers to the **V**ery **H**igh **S**peed **I**ntegrated Circuit program. The Department of Defense (DoD) sponsored this program with the goals of developing a new generation of high-speed integrated circuits. During the course of the program, the increasing complexity of digital systems that were made possible by continuous advances in semiconductor and packaging technologies was found to have a fundamental impact on the economics of the design of military and space electronic systems. When the life cycle costs of these systems were examined, it was evident that the cost of maintenance was becoming significant. Furthermore, it was becoming increasingly difficult to share designs of subsystems across contractors. The need for a standardized representation of digital systems became apparent. A team of DoD contractors was awarded the contract to develop the language, and the first version was released in 1985. The language was subsequently transferred to the IEEE for standardization, after which representatives from industry, government, and academe were involved in its further development. Subsequently, the language was ratified in 1987 and became the IEEE 1076–1987 standard. The language was reballoted after five years and, with the addition of new features, forms the 1076–1993 version of the language. This text adheres to the 1076–1993 standard while attempting to point out differences with the 1076–1987 version, as one is bound to encounter legacy models. Unless otherwise stated, simulation and synthesis compilers will generally support both versions and will provide mechanisms to control which version of the language is being supported.

Ever since VHDL became an IEEE standard, it has enjoyed steadily increasing adoption throughout the electronic systems computer-aided design (CAD) community. The DoD requires that VHDL descriptions be delivered for all application-specific integrated circuits (ASICs). The establishment of the IEEE 1164 standard package has improved the interoperability among models developed using VHDL environments from different CAD vendors. Every major CAD vendor supports VHDL. This status of VHDL as an industry standard provides a number of practical benefits, including model interoperability among vendors, third party vendor support, and design reuse.

Conventional procedural programming languages, such as C or Pascal, describe procedures for computing a mathematical function or manipulating data—for example, matrix multiplication or sorting, respectively. In these languages, a program is a recipe consisting of a sequence of steps defining *how* to perform a computation or manipulate data. The execution of the program results in the computation or rearrangement of data values. On the other hand, VHDL is a language that *describes* digital systems. A simulator will use its descriptions to simulate the behavior of the system without having to actually construct it. Alternatively, synthesis compilers can utilize such a description to create descriptions of the digital hardware for implementing the system. Although VHDL has been investigated for its use in describing and simulating analog systems, the language is used predominantly in the design of digital electronic systems.

1.2   Digital System Design

The design of digital systems is a process that starts from the specification of requirements and proceeds to produce a functional design that is eventually refined through a sequence of steps to a physical implementation. Both simulation and synthesis are complementary activities employed in the design process. Consider the design of an application-specific integrated circuit (ASIC) for processing digital images. This is a custom chip designed for a specific task, unlike a microprocessor that may be programmed for a variety of tasks. Custom ASICs are generally the highest performing solution for any computation. Figure 1.1 shows an example of the sequence of activities that typically takes place during ASIC design.

The first step is the specification of the requirements that the chip is to satisfy. Such a specification will typically include the performance requirements derived from the number of images to be processed/sec and the operations to be performed on them, as well as interface requirements, cost constraints, and other physical requirements, such as system size and power dissipation. From these functional requirements, one can generate a preliminary high-level functional design. Simulation is often used at this level to converge to a functional design that can meet the specified performance requirements. This initial functional design is now refined to produce a more detailed design description at the level of registers, memories, arithmetic units, and state machines. This is the register transfer level (RTL) of the design. Subsequent refinement of this RTL description produces a logic design that implements each of the RTL

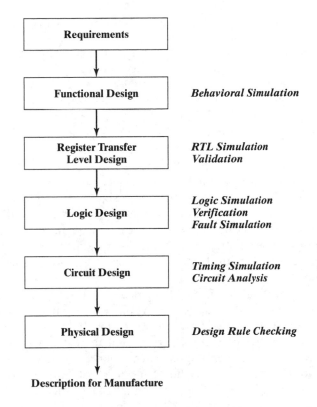

FIGURE 1.1 Typical activity flow in top-down digital system design

components. Both RTL and logic level simulation then can ensure that the design meets the original specification. Fault simulation can model the effects of expected manufacturing defects, as well as faults that may be induced due to the environment. For example, if this image-processing chip is to be flown in a satellite, radiation effects in space can cause devices to change state and can lead to single-bit errors. If the error rate in the intended orbit is relatively high, the designer can modify the design to accommodate such bit errors using techniques tuned to the particular physical phenomena. Finally, the designer may transform the logic level implementation into a circuit level implementation and thence to a physical chip layout, from which the designer can evaluate accurate physical properties of the design, such as chip area and power dissipation. Design rule checks, circuit parameter extraction, and circuit simulation activities can be performed at this level.

At each level of this design hierarchy, various components describe the design. At the higher or more abstract levels, we have a smaller number of more powerful components such as adders and memories. At the lower and less abstract levels, we have a larger number of simpler, less powerful components, such as gates and transistors. Each level of the design hierarchy corresponds to a *level of abstraction* and has an associated set of activities and design tools that support the activities at this level. Figure 1.1 shows some of the activities at each level. The accuracy with which we can predict the behavior, physical properties, and performance of the circuit increases at the lower levels of the hierarchy with considerably longer simulation times. Imagine having to simulate the behavior of 1 billion transistors on a chip!

If design errors are discovered at these finer levels of detail, changes in the design may be expensive to make, particularly if we have to move back several levels in the design process to correct these errors. This can lead to longer development times and, consequently, increased cost, not to mention loss of revenues by being late to market. Much of the motivation for the development of hardware description languages in general stems from the evolving economics of the marketplace for electronic systems and the methodologies used to design these systems. With the ability to simulate designs at multiple levels of abstraction, errors can be discovered and corrected early. Moreover, it is important to note that, throughout this hierarchy, simulation is a commonly used technique. Hardware description languages such as VHDL are targeted for use throughout this design hierarchy and provide some degree of uniformity across the various levels.

By facilitating the automation of design processes, VHDL-based tools reduce design costs. The cost of designing ASICs today can vary from $5M to $100M. For example, an ASIC with 80M transistors in 90-nm technology for a leading-edge embedded systems application is projected to cost upwards of $85M [9]. Such cost structures mandate a high degree of confidence in the design and an increasing reliance on tools for automating the design processes, thereby reducing the costs associated with design. Hardware description languages such as VHDL are at the center of such automation strategies. More on the impact of cost structures will be presented in the next section.

1.3   The Marketplace

The use of computer-aided design environments in general, and hardware description languages in particular, is driven by the technology marketplace. After all, these tools and languages are intended to enable the cost-effective and profitable development of electronics products. The key question with respect to the state of the art is, Where are the costs, both in terms of costs incurred and revenues lost, due to design paradigms that are not as efficient or effective as they could be?

The costs incurred are directly a function of the available technology. Both the semiconductor industry and the software industry have been in overdrive during the last four decades and show no signs of abatement in the next decade. Memory densities have been quadrupling every three years and processor speeds have been doubling every 18 months. As chip densities increase, new products are becoming available at faster rates and development cycles are becoming shorter. Madisetti captures the importance of time to market in an illustrative manner in Figure 1.2 [12]. For every product, there is an optimal time for its introduction into the marketplace in a manner that will maximize the revenue generated. Early in the life cycle of the product,

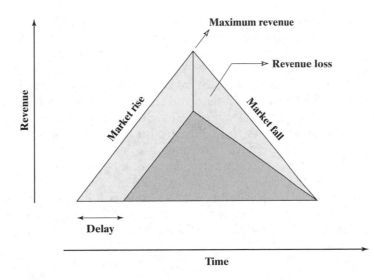

FIGURE 1.2   Business model for revenue as a function of time to market [12]

market share should rise, peak, and then begin to decline as newer products emerge. If the product is delayed to market, then the trajectory of its performance relative to the maximum may be depicted as shown in the figure. In sum, the figure shows that it is very difficult to make up for lost ground. Many organizations measure lost opportunity cost in terms of thousands to tens of thousands of dollars per day in lost revenue. An unfortunate side effect of such economics is that time-to-market pressures may reduce the quality of the first version of the product. With the increasingly faster rate of technology evolution, approaches to reduce time to market are critical. New CAD tool environments and hardware description languages are an integral part of any solution.

Given this emphasis on the time to market and the resulting need for shorter design cycles, what are the impediments to shortening the time to market? Historically, the CAD tool industry has developed in a bottom-up fashion and has been driven by the development of point tools and processes: design rule checking, layout, verification, and so forth. Digital design was a manual process in that designs needed to be transformed manually between levels of abstraction and, at each level of abstraction, we would be assisted by the appropriate tools. This is partly a by-product of the fact that, for chip and board designs, there has been an understanding of the specific design problems, which include layout, design rule checking, test vector generation, and so on. Advances have focused on better tools to address these problems, spurred by the new challenges of faster, denser, more complex technologies. In contrast, very few tools exist at the architectural level for addressing system design issues. The problems are not as well defined, but the impact can be overwhelming. Observers note that the first 10%–20% of the design cycle can determine 70%–80% of the final system cost. Moreover, reports indicate that only 5%–10% of the design cycle time is spent in studying and formulating requirements, while customer requirements affect 70% of the manufacturing costs [12]. Furthermore, costs are rapidly increasing with each new generation of technology. One study reports that designing an 80M transistor chip using 0.1-micron feature size technology for a leading embedded systems application will cost on the order of $80M–$90M [9]. Such cost structures make the role of auto-mated CAD tools ever more critical in reducing design costs, increasing the importance of verification at design time, and ensuring that performance specifications of the resulting design are met. Less aggressive designs cost less, but can still cost upward of $5M–$10M for a new design, and these costs are increasing with each new generation. Furthermore, time-to-market pressures are increasing to maintain the competitive edge. All of these factors further underscore the importance of hardware description languages and the design tools structured around them.

New methodologies are needed to address the issues of requirements capture, system specification, and early analysis via rapid prototyping. Such new design methodologies are emerging, and VHDL is becoming an integral part of such design approaches.

1.4   The Role of Hardware Description Languages

Traditional design methodologies have been structured around a hierarchy of representations of the system being designed. Distinct representations at differing levels of detail are necessary for the various tasks encountered during design. One of the best-known representations of the different views and levels of abstraction in a digital system is the Y–chart shown in Figure 1.3 [6, 18, 19] and illustrated through the next example.

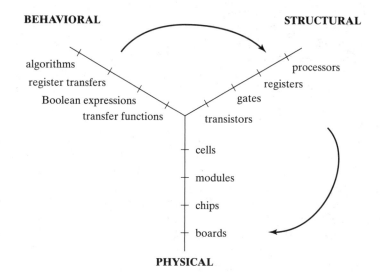

FIGURE 1.3   Design views and corresponding levels of abstraction

Imagine a company, DSP Inc., in the business of designing a next-generation digital signal-processing chip, code name Cyclone. Given the current costs of fabrication and the time window within which the chip must be brought to market to be competitive, we wish to verify prior to fabrication that the chip can support the intended applications. From the design flow shown in Figure 1.1, we see that we can describe this behavior at multiple levels of abstraction—that is, at the functional level, RTL level, logic level, and so on. Early in the design, behavioral descriptions are necessary so that simulations can ensure that the chip is functionally correct. This functionality may be verified at more than one level of abstraction. The advantage of this approach is that we can make this assessment independent of the many possible physical implementations. Once we have verified the functionality, we can translate the design to a structural description made up of the major components of this chip: memories, registers, arithmetic units, logic units, and so forth. We can employ simulation again to ensure that the Cyclone structural design correctly performs the intended functions using the components that we have selected. As shown in Figure 1.3, the structural description may also include varying levels of detail. The design can be refined until we translate this description to a physical description that we can further refine to produce a manufacturable specification.

Historically, hardware description languages have been targeted to a certain level of abstraction, such as the gate level or register transfer level. Tools have been developed and targeted for tasks at a specific level of the design. For example, some tools are optimized for implementing state machines that describe data movement at the register transfer level. Other tools have been developed for verification at the gate level by generating test vectors used for final chip verification. The tools at the different levels of abstraction may employ different descriptions of the Cyclone. Such *point tools* are focused on a single aspect of the design and on a single level of abstraction. Often, these tools come with their own languages and associated compilers and simulators. However, as chips become more complex and design processes start using an increasingly diverse set of tools, the time taken to move information between tools has become a concern. Recently, the lost productivity due to incompatibility between design tools has been estimated to be as high as $4.5 billion dollars/year [12]. While simulation has been used at all levels of system design designated in Figure 1.3, synthesis attempts to move in an automated fashion among the three domains and levels of abstraction. The use of hardware description languages such as VHDL can help address several aspects of this fundamental problem of design refinement.

Interoperability => The VHDL language provides a set of constructs that one can apply at multiple levels of abstraction and multiple views of the system. This significantly expands the scope of the application of the language and promotes a standardized, portable model of electronic systems. Thus, the complexity of the movement of data and design information between tools is significantly reduced. The net effect is a reduction in the time to design the Cyclone and bring it to market. We may also expect that the reduction in the number of distinct types of tools and languages realizes a reduction in the cost of the design infrastructure and hence a reduction in the product unit cost.

Technology is a rapidly moving target. We cannot anticipate innovations in technology design styles or products. Therefore, an attractive philosophy is to develop a design environment that is independent of technology. The industry is characterized by a number of distinct technologies and associated design styles that target specific points in the continuum of time to market, cost, and performance. For example, the use of programmable logic devices (PLDs) such as Field Programmable Gate Arrays (FPGAs) has the attributes of low cost and quick time to market. ASIC products incur higher nonrecurring engineering development costs. However, they can deliver substantially higher performance and, in volume, can lead to lower costs and lower power consumption. These distinct design styles lead to environments with distinct sets of CAD tools and methodologies.

Technology Independence => The VHDL descriptions of the design are not tied to a specific methodology or target technology. The language is rich enough that it can be used to describe a chip at the instruction set level, register transfer level, or switching transistor level. DSP Inc. must produce a high-performance version for use in a real-time radar processing system for the military. Design tools for custom and ASIC chips utilize VHDL in their suites to simulate and validate their designs prior to detailed physical design. Prior to physical design, we would wish to have detailed timing information to ensure that the performance requirements of the application can be met. However, the application software developers may simply want a functionally accurate instruction set simulation of the Cyclone chip so that development of application software may begin immediately—hardware/software codesign! Eventually, when a detailed physical design is available, the software can be tested on an associated simulation to determine the performance that can be achieved for the applications of interest—two widely differing levels of detail, both easily supported within the same language.

Now, six months later, you have found a commercial market for the Cyclone to implement video compression algorithms. The only problem is that they have to cost a tenth of the cost of the military part, but the processing constraints are much looser. The major CAD vendors have tools that can synthesize designs to FPGA and complex PLD (CPLD) devices. The Cyclone is reimplemented within an FPGA using synthesis from the VHDL models. The result is a cheaper, slower, higher volume product.

Now let us say that your company develops a model of the Cyclone using CAD tools from vendor Tools'R'Us. Now your company wishes to have a subcontractor use this chip in the design of a second product, say a voice recognition board for personal communicators. Their design environment is completely different, using archaic design tools from StoneAge CAD Tools, Inc. You now need to send them detailed schematics and operational specifications and educate them on the Cyclone design so that they may use this description to design and validate the board-level product using their design tools. However, they do support the VHDL language. Since VHDL is a standard, rather than having them reconstruct the design in their environment, you can use and simulate the preceding instruction set models of the Cyclone. They will produce identical results, using StoneAge's simulators.

Design Reuse => We can see libraries of VHDL models of components emerging and being shared across platforms, toolsets, organizations, and technical groups. Engineers working on a large design can be independently designing subsystems with considerably less concern for design environment or design tool compatibility issues. As we will see in later chapters, the language possesses many features that enable us to separate the interface of a component from its internal implementation.

Hardware/Software Prototyping => Part of the difficulty with a board design is that the software for processing the data streams cannot be tested until hardware is available. But what if we have detailed hardware descriptions of the components on the board that behave exactly as the Cyclone chip does, to the level of detail of a clock cycle? We could simulate the system to a level of detail that would permit the simulator to take the place of the hardware for software development purposes. A software simulation could execute application programs, permitting trade-offs between the hardware and software implementations, even before a single chip or board was designed! Such an approach based on this concept of Virtual Prototyping is an example of a design methodology for systems in general, but it is currently being applied to digital signal processing systems in particular [12, 15]. The use of VHDL at multiple levels of abstraction is illustrated in Figure 1.4.

Hardware description languages are at the core of modern digital systems design methodologies. Proficiency in their use will be necessary for designers, and familiarity will be a must for electrical and computer engineering students.

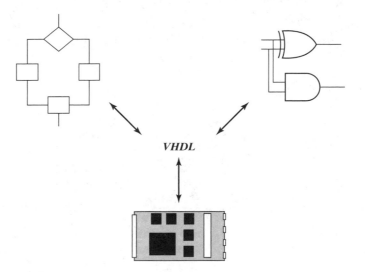

VHDL

FIGURE 1.4   The VHDL language can be applied at multiple levels of abstraction

1.5   Chapter Summary

Digital system design is a process of creating and managing multiple descriptions of systems representing distinct views and varying levels of abstraction. Historically, design environments grew around point tools for solving specific problems at each level of abstraction. Design methodologies governed the application of these tools and the flow of information between them. The evolution of distinct methodologies and modeling styles can hinder interoperability and make it difficult to share models. As electronic systems have grown in complexity, the need to integrate these point tools into a cohesive design process has determined a large component of the economics of product design. The advent of hardware description languages such as VHDL and their acceptance as industry standards has had an enormous impact on the economics of product development.

We see that VHDL modeling can be used to model hardware and software systems at multiple levels of abstraction. The language is independent of technology and design methodologies or styles and therefore promotes portable descriptions, rapid prototyping, and the free exchange of models among organizations and individuals. The result has been the promise of reduced design cycle times, faster time to market, and reduced cost. We can expect to see the use and growth of the language continue as a widely used hardware description language for both military and commercial systems for some time to come.

Within the purview of the design of electronic systems, several other hardware description languages accompany the VHDL language. One of the more widely used is Verilog. The goals and motivation for the Verilog language parallel those of VHDL, although Verilog has a distinct developmental heritage. The major CAD tool vendors support both the VHDL and Verilog languages, as well as a number of homegrown and specialized languages for specific design activities.

CHAPTER 2

Modeling Digital Systems

VHDL programs are unlike programs written in Java, C, or Fortran. Conventional programs are based on thinking in terms of algorithmic sequences of calculations that manipulate data toward a specific computational goal. The thought process that goes into writing such programs is inherently procedural, a direct result of the serial computing model embodied in most modern computers. Writing VHDL programs is very different. We are not interested so much in *how* a function is computed; rather we are interested in describing the behavior of some physical system such as a digital circuit. This behavioral description can be used for at least two purposes. The first is the simulation of digital circuits. A simulator uses the VHDL description to conduct a simulation that "behaves like" the physical system. Such simulations can verify the behavior of the digital circuit prior to expensive and time-consuming fabrication. The simulation can, in fact, serve as a virtual prototype in making and evaluating design trade-offs prior to finalizing the hardware design. The second purpose is for the synthesis of digital circuits. Design tools analyze the VHDL description and produce a digital circuit that implements the behavior captured in the VHDL description. The resulting circuit descriptions can be processed rapidly to produce custom hardware or can be used to configure reprogrammable hardware components to implement the design. Thus, the designer can, in fact, use the VHDL descriptions to support two complementary processes found in the design of digital systems: simulation and synthesis.

The core motivation for the VHDL language was to describe digital systems. A natural question to ask is, What aspects of digital systems should be captured to provide a complete (in the sense required for simulation or synthesis) description? Equivalently, what aspects should be ignored? This chapter discusses the significant structural, physical, and behavioral characteristics of digital systems. It is these characteristics that the VHDL language constructs described in the following chapters capture. Thus, the main language constructs follow naturally from attributes of digital systems. A basic knowledge of digital systems allows one to quickly come up to speed in the use of the language.

2.1   Motivation

Physical systems are characterized by interactions among potentially thousands to millions of *concurrently* operational components. A description must be able to capture the fact that several activities, such as the operation of a gate, multiplexor, or adder, are taking place concurrently in time. Sequential descriptions captured in conventional programming languages are not naturally suited for this purpose. They describe *how* to compute a result, not *what* the computing entity is. If we can accurately describe the hardware design, and not necessarily what it computes, we can use this description to determine whether the design is correct. We can check a description of an adder by simulating the hardware described by the VHDL program, providing a sequence of input values, and checking the corresponding output values. For example, consider describing a modern microprocessor at the gate level, which at the time of this writing can be represented by Intel's Itanium. If we could formally prove the correctness of a design from the design specification and implementation, we would not need simulation. However, formal proof techniques cannot currently, nor in the foreseeable future, handle designs of such complexity. Thus, we rely on simulation to establish the correctness of the design. We also often rely on simulation to evaluate the performance of a design. Frequently, the reason we want to simulate a physical system is that we cannot otherwise evaluate the system. For example, if we need to know the average time a person must wait at a bus stop, we cannot obtain some mathematical function that will accurately compute this value for us. Traffic patterns, whimsical pedestrians, and riders with incorrect change all contribute to a degree of unpredictability and complexity that prevents us from writing an accurate mathematical expression for the waiting time at a bus stop. However, we can simulate the transit system and observe how long people wait in the simulation. If our simulation of the individual activities is accurate, then we will be able to predict reliably the delays that will occur in practice.

The digital systems that we will consider do not exhibit probabilistic or random behavior but are composed of many constituent subsystems whose interactions can be quite complex and infeasible to analyze analytically. Therefore, we may simulate the design prior to implementation to ensure that the system meets its performance specification. For example, we may design a single-chip media processor that interfaces to a camera and processes images in real time. We can then use a VHDL simulation of the chip to verify that the design can indeed operate fast enough to keep up with the rate at which images are being received from the camera. Given the cost of modern fabrication facilities and the increasing complexity of digital systems, it has become necessary to be able to rely on accurate simulation models to design and test chips and systems prior to their construction. How can we be sure that the design will function as intended or that the design is indeed correct? The VHDL simulation serves as a basis for testing complex designs and validating the design prior to fabrication. The overall effect is that of reducing redesign, shortening the design cycle, reducing the probability of design error, and bringing the product to market sooner.

The features of a language for describing digital systems are quite different from those of procedural languages. While early simulators for digital systems have been written in C, C++, or Java, the developers had to provide new functions, operations, and data types to enable one to write simulation applications. The definition of the VHDL language provides a range of built-in features in support of the simulation of digital systems.

As a result of the motivation to model digital systems, many of the language concepts and constructs can be identified with the structural, behavioral, and physical characteristics of digital systems. We learn best when we can identify with concepts with which we are already familiar. In this chapter, we will review the operational characteristics of digital systems and identify several key attributes that will subsequently be linked to major language features. The chapter concludes with a description of the discrete event simulation model that underlies the execution of VHDL programs. The remainder of the text focuses on the presentation, description, and application of the major language features.

2.2   Describing Systems

The term *system* is used in many different contexts to refer to anything from single chips to large supercomputers. Webster's dictionary defines a system as "an assemblage of objects united by some form of regular interaction or interdependence."

We are interested in being able to describe digital systems at any one of several levels of abstraction, from the switched transistor level to the computing system level. To do so requires us to identify attributes of systems common to all of these levels of abstraction. For example, imagine that you are in the business of selling sound cards for personal computers. You are trying to make a sale to Personal Computer, Inc., and have the company include this sound card as a part of its product line, making the card available in all of its personal computers. A sample design of your card might appear as shown in Figure 2.1. In order to make this sale, you must be able to provide a concise description of this card to the engineers of Personal Computer, Inc. How could you describe such a card? What do the engineers need to know to evaluate your design? They certainly need to understand the *interface* to the card. For example, what can you connect to this card? Speakers, microphones, or even a stereo amplifier? How does the processor communicate with this card? You must be able to describe all of the signals that may pass through the card interface to ensure compatibility with the internal bus used for communication between components within their system. A second aspect of this description is the behavior of the card itself. This could be communicated in one of several ways. One way is to describe component chips and their interconnection, assuming that the engineers are familiar with the operation of the individual chips. Such a description is commonly referred to as a *structural* description and can

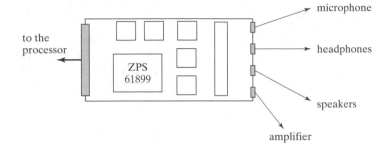

FIGURE 2.1   An example of a system

easily be conveyed in a block diagram. Alternatively, we can describe the behavior of the card in terms of the type of processing it performs on the input signals and the type of output signals it produces—for example, audio output for the speakers. Such descriptions are referred to as *behavioral* descriptions. You are describing what the card does, independently of the physical parts that make up the card. Note one important point: The behavioral description does not say anything about how the card is built. For example, I can get the same behavior at different costs (but perhaps with better quality). Depending upon who you are talking to, one description or the other is preferable. For example, marketing would be interested in the behavioral description and engineering would be interested in the structural description.

Structure and behavior are complementary ways of describing systems. The specification of the behavior does not necessarily tell you anything about the structure of the system or the components used to build it. In fact, there are usually many different ways in which you can build a system to provide the same behavior. In this case, other factors such as cost or reliability become the determining factors in choosing the best design. We would expect that any language for describing digital systems would support both structural and behavioral descriptions. We would also expect that the language would enable us to evaluate or simulate several structural realizations of the same behavioral description. The VHDL language does indeed provide these features.

2.3   Events, Propagation Delays, and Concurrency

Let us look a little closer at these structural and behavioral descriptions. Digital systems are fundamentally about *signals*—specifically, binary signals that take the value 0 or 1. Digital circuits are made up of *components* such as gates, flip-flops, and counters. Components are interconnected by wires and transform input signals into output signals. A machine-readable (i.e., programming language) description of a digital circuit must be able to describe the components that make up the circuit, their interconnection, and the behavior of each of the components in terms of its input and output signals and the relation between them. This language description can then be simulated by associated computer-aided design tools or used to synthesize a hardware description.

Consider the gate level description of a half adder shown in Figure 2.2. There are two input signals, a and b. The circuit computes the value of two output signals, sum and carry. The values of the output signals, sum and carry, are computed as a function of the input signals, a and b. For example, when a = 1 and b = 0, we have sum = 1 and carry = 0. Now, suppose the value of b changes to 1. We say that an *event* occurs on signal b. The event is the change in the value of the input signal from 0 to 1. In our idealized model of the world, this transition takes place instantaneously at a specific, or discrete, point in time. Real circuits take a finite amount of time to switch states, but this approximation is still very useful. From the truth tables for the gates, we know that such an event on b will cause the values of the output signals to change. The signals sum and carry will acquire values 0 and 1, respectively; that is, events will occur on the signals sum and carry. A basic question then becomes the following: When will these events on the output signals occur relative to the timing of the events on the input signals?

Electrical circuits have a certain amount of inertia or natural resistance to change. Physical devices such as transistors that implement the gate level logic take a finite amount of time to switch between logic levels. Therefore, a change in the value

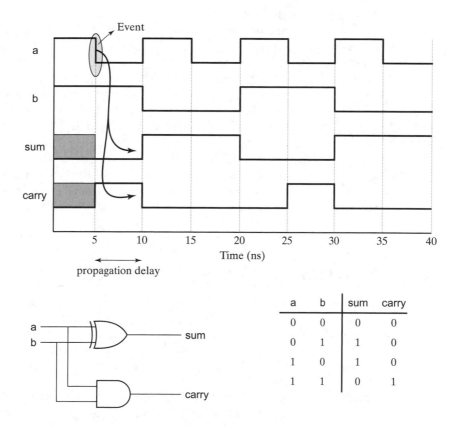

FIGURE 2.2 Half-adder circuit

of a signal on the input to a gate will not produce an immediate change in the value of the output signal. Rather, it will take a finite amount of time for changes in the inputs to a gate to propagate to the output. This period of time is referred to as the *propagation delay* of the gate. The time it takes for changes to propagate through the gate is a function of the physical properties of the gate, including the implementation technology, the design of the gate from basic transistors, and the power supplied to the circuit. From the timing behavior depicted in Figure 2.2, we can see that the gate propagation delay is 5 ns. Electrical currents that carry logic signals through interconnect media such as the wires also travel at a finite rate. Thus, in reality, signals experience propagation delays through wires, and the magnitude of the delay is dependent upon physical properties of the interconnect such as the length of the wire. This delay is nonnegligible, particularly in very high-speed, high-density circuits. It is interesting to note that wires have considerably less inertia than gates. The resulting physical phenomenon is therefore quite different. However, as device feature sizes have become increasingly smaller, wire delays have become nonnegligible in modern high-density circuits. As we shall see in later chapters, VHDL provides specific constructs for handling both types of delays. The timing diagram shown in Figure 2.2 does not include wire delays.

A third property of the behavior of the circuit shown in Figure 2.2 is *concurrency* of operation. Once a change is observed on input signal b, the two gates concurrently compute the values of the output signals sum and carry, and new events may subsequently occur on these signals. If both gates exhibit the same propagation delays, then the new events on sum and carry will occur simultaneously. These new events may go on to initiate the computation of other events in other parts of the circuit. For example, consider two half adders combined to form a full adder, as shown in Figure 2.3. Events on the input signals In1 or In2 produce events on signals s1 or s3. Events on s1 or s3 in turn may produce events on s2, sum, or c_out. In effect, events on the input signals In1 or In2 propagate to the outputs of the full adder. In the process, many other events internal to the circuit may be generated. In the associated timing diagram, every $0 \rightarrow 1$ and $1 \rightarrow 0$ transition on each signal corresponds to an event. Note the data-driven nature of these systems. Events on signals lead to computations that may generate events on other signals.

2.4   Waveforms and Timing

Over a period of time, the sequence of events that occur on a signal produces a *waveform* on that signal. The effects of each event may in turn propagate through the circuit, producing waveforms on internal signals and eventually producing waveforms on the output signals. The timing diagram shown in Figure 2.3 is a collection of waveforms on signals in the full-adder circuit, where each waveform is an alternating sequence of $0 \rightarrow 1$ and $1 \rightarrow 0$ transitions or events.

The model of the operation of digital circuits in terms of events, delays, concurrent operation, and waveforms extends to sequential circuits as well as combinational circuits. Consider the operation and timing of a positive edge-triggered D flip-flop

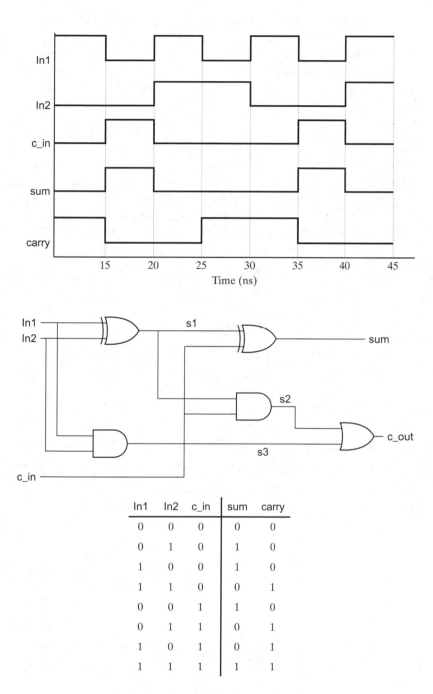

FIGURE 2.3   Full-adder circuit and truth table

shown in Figure 2.4. The output values are determined at the time of a $0 \rightarrow 1$ transition on the clock signal. At this time the input value on signal D is sampled and the values of Q and $\overline{Q}$ are determined. Events on the asynchronous set ($\overline{S}$) and reset ($\overline{R}$) lines produce events on the output signals independent of events on the clock. The unique aspect of the behavior of this model is the dependency on the clock signal. The computation of output events begins at a specific point in time determined by a $0 \rightarrow 1$ event on the Clk signal, independent of events occurring on the D input signal. This need to *wait for* a specific event is an important aspect of the behavior of sequential digital circuits. Such circuits are referred to as *synchronous* circuits. Synchronous circuits operate with a periodic signal commonly referred to as a clock that serves as a common time base. Clocks are an important aspect of digital circuits and deserve special attention.

Alternatively, in the absence of a global clock signal, many digital systems operate asynchronously with request–acknowledge protocols. This is most easily understood in the context of a communication channel between two chips. The transmitting

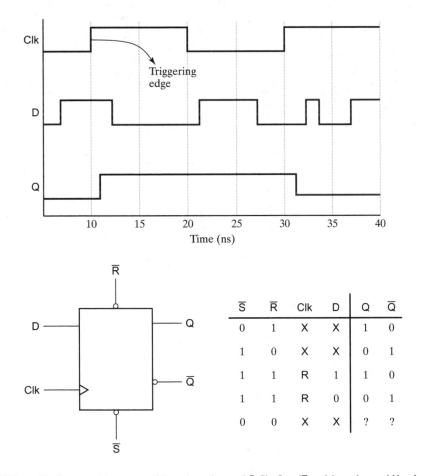

FIGURE 2.4   Excitation table for a positive edge-triggered D flip-flop (R = rising edge and X = don't care)

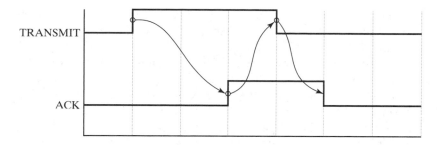

FIGURE 2.5   An example of a four-phase handshake

chip may assert a TRANSMIT signal when it is ready to transmit data. The receiving chip will be monitoring the TRANSMIT signal, and when the signal is asserted, the chip will read the data and then assert an ACK signal. On seeing the event on the ACK signal the transmitter can stop transmitting the data. The timing is shown in Figure 2.5, where the arrows represent the causal relationship between signal transitions. Such "handshaking" operations are very common in the operation of digital logic implementation of communication channels, and it is clear that both the transmitter and receiver have to wait for specific events on a signal.

2.5   Signal Values

Signal values are normally associated with the outputs of gates. Wires transfer these values to the inputs of other gates, which, as a result, may drive their outputs to new values. In general, we tend to think of signals in digital circuits as being binary valued and being driven to these values by a source such as the power supply or the output of a gate. These logical values are physically realized within a circuit by associating logical 0 or 1 values to voltage or current levels at the output of a device. For example, some circuits recognize a voltage occurring between 0 and 0.8 volts as a logical 0 signal, and a voltage occurring in the range 2.0 to 3.3 volts as a logical 1 signal. However, what happens when a signal is not driven to any value—for example, if it is disconnected? What is the value of the signal? It is neither 0 nor 1. Such a state is referred to as the high impedance state and is usually denoted by Z. This is a normal, inactive condition and occurs when a signal is (temporarily) disconnected.

What happens when a signal is concurrently driven to both a 0 and a 1 value? This is clearly an abnormal or error condition and should not occur. It is indicative of a design error. How do we denote the value of the signal? Remember that our overall goal is the accurate description of digital systems, often for the purpose of testing a design to ensure that it is correct prior to fabrication. If this condition were to occur during the simulation of a circuit, the simulator must be able to represent the value of the signal and propagate the effects of this design error through the circuit. Such unknown values are typically denoted by X. What if the initial value of a signal is undefined? How can we represent this value and propagate the effects of uninitialized signal values to determine the effect on the operation of the circuit? Such values are typically denoted by U.

At the very least, we see that for the purpose of simulation, 0 and 1 values alone are insufficient to capture the behavior of digital systems accurately. We will see that the VHDL language is flexible enough to enable the definition of a range of values for single-bit signals. Early in the evolution of VHDL, CAD tool vendors defined their own value systems. Some vendors even had as many as 46 distinct values for a single-bit signal! This made it difficult to share VHDL models. Imagine if different C compilers had different definitions of the values of integers. The same C program could produce different results, depending upon the compilers that were used. In addition to 0, 1, Z, X, and U values, it is useful to denote the concept of *signal strength*. The strength of a signal reflects the ability of the source device to supply energy to drive the signal. This strength can be weakened or attenuated by, for example, the resistance of the wires giving rise to signals of different strengths. Although the range of strength values can be large, only two levels are sufficient to characterize many types of transistor circuits. The use of strength values facilitates certain styles of design. One common use of the VHDL language is to describe the behavior of circuits that can automatically be synthesized by design tools. A value system that incorporates the concept of signal strength is therefore necessary.

To establish common ground and enable the construction of portable models, the IEEE has approved a 9-value system. This is the IEEE 1164 standard, which has gained acceptance and widespread usage. In this system, single-bit signals take on functional values of 0, 1. However, they can also be unknown (X), uninitialized (U), or not driven (Z). It also includes two levels of signal strength and the don't-care value (-), as shown in Figure 2.6. It is important to note that this value system is not a part of the VHDL language, but is rather a standard definition that vendors are motivated to support and users are motivated to use, since it enables reuse of designs and sharing of models among users. Practically all vendors support the IEEE 1164 value system.

Value	Interpretation
U	Uninitialized
X	Forcing Unknown
0	Forcing 0
1	Forcing 1
Z	High Impedance
W	Weak Unknown
L	Weak 0
H	Weak 1
-	Don't Care

FIGURE 2.6   IEEE 1164 value system

From the perspective of the synthesis of digital circuits from VHDL descriptions, the effect of the value system is a bit different. For example, when circuits are synthesized, the unknown and uninitialized values do not have any meaning for a signal. A signal must be represented by a wire or a storage element such as a latch or flip-flop, and the signal value is represented by the state of the wire, latch, or flip-flop. There is no physical implementation for the unknown or uninitialized signal values. Synthesis compilers must address this issue.

2.6 Shared Signals

It is common for components in a digital circuit to have multiple sources for the value of an input signal and multiple destinations for the value of an output signal. However, connecting all component inputs and outputs by dedicated signal paths can be expensive. Therefore, many designs will utilize *buses*: a group of signals that can be time shared among multiple source and destination components. Consider the simple case of a single-bit signal shared among multiple sources and destinations, as shown in Figure 2.7. This is an efficient design in the sense that it minimizes the interconnect wiring among communicating components. Sources and destinations time share the bus using transceivers—transmitter/receiver pairs. At any given time, only one transmitter is enabled while multiple receivers may be enabled. A decoder can control the tri-state buffers to ensure that only one source is driving the bus at a time. The architecture of personal computers and workstations is built around one or more multibit buses. The microprocessor may drive a bus to communicate addresses to memory, while, at other times, the memory controller may drive a bus to return values to the processor. The Input/Output buses in PCs interconnect many devices, such as CD ROMs, floppy disk drives, and hard disk drives, that share the Input/Output bus. Thus, it is important to be able to represent signals with multiple possible *drivers*.

While buses permit only one transmitter at a time, certain forms of switching circuits have a design based on *wired logic* and permit multiple concurrent transmitters or drivers on a bus. In these circuits, the interconnection of wires can produce AND and OR Boolean functions. For example, if several devices drive a shared signal to either 0 or Z, then the signal value will be determined by the interactions among all of the values applied to it: In this case, if at least one device is driving the signal to a 0, the value of the signal will be 0. Thus, the interconnection can be regarded as implementing a wired-OR function. Similarly, it is also possible to implement wired-AND logic.

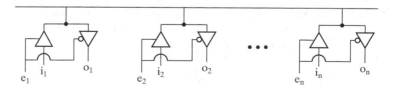

FIGURE 2.7 An example of a signal driven by multiple sources—buses

The key issue here is that some signals may be driven to a value by one or more sources rather than by a single source, such as the output of a gate. Hardware description languages must be expressive enough to describe such circuits for the purpose of accurate simulation. Language constructs must be capable of capturing the interaction among multiple drivers for a signal. For example, is the resulting value of the signal the AND, OR, or maximum value of all of the drivers?

Similarly, synthesis compilers must be able to process such descriptions and produce logic that arbitrates among accesses to shared signals in a manner that is correct. For example, if only one source is permitted, then a decoder must be synthesized to control the multiple drivers.

2.7   Simulating Hardware Descriptions

For simulation we start with the specification of the behavior of a digital system and we construct a VHDL model of this system. This description of a hardware design captures the attributes of the design described in the preceding sections. For example, we describe events on the input signals to the design, how values for output signals are computed, and a specification of when these values drive output signals. A VHDL simulator executes this model to mimic the behavior of the physical circuit—the occurrence events and waveforms on signals. How is this description used to drive an accurate simulation of the hardware? If we are to understand how to utilize hardware description languages effectively we must be comfortable with the underlying models for their execution. This section addresses the execution model underlying the use of VHDL for simulation.

In contrast, digital circuit synthesis is the reverse process. A VHDL program is the input to a synthesis compiler that can process this description to generate the physical design of a circuit. The compiler must infer the hardware structures necessary to implement the behavior described by the VHDL code. Essentially, the synthesis compiler mimics the activities of a human designer who generates a hardware design from an initial specification. However, automated synthesis, when feasible, is much faster than a designer, contributing to the reduction of the overall product design time and product cost. Figure 2.8 captures the relationship between simulation and synthesis. The VHDL model provides a description of the behavior of a digital circuit. From the perspective of simulation, this model is used to study the properties of the circuit. In a sense, a knowledge of the physical and behavioral properties of the circuit lead to the model. In synthesis, the model is used as the first step in generating a physical design. This text focuses on the use of VHDL for simulation. A companion text, *Introductory VHDL: From Simulation to Synthesis,* that also addresses synthesis with an approach similar to that found in this text is [20] while there several other texts that deal with VHDL synthesis (some examples include *A VHDL Synthesis Primer* [3] and *HDL Chip Design: A Practical Guide for Designing, Synthesizing, and Simulating ASICs and FPGAs Using VHDL or Verilog* [17]).

Just as we reason about C++ programs using the serial computing model as a basis, we need an understanding of how VHDL programs execute in order to reason

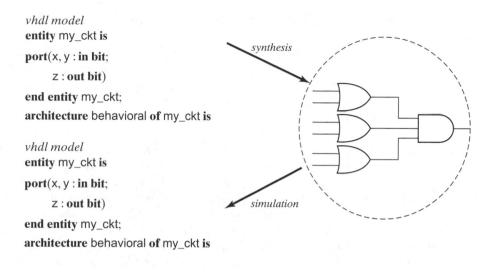

vhdl model
entity my_ckt **is**

port(x, y : **in bit**;

z : **out bit**)

end entity my_ckt;
architecture behavioral **of** my_ckt **is**

vhdl model
entity my_ckt **is**

port(x, y : **in bit**;

z : **out bit**)

end entity my_ckt;
architecture behavioral **of** my_ckt **is**

FIGURE 2.8   Simulation vs. synthesis

about their performance and correctness. While the serial computation model has been prescribed for mechanical computation, the VHDL execution model focuses on mimicking the occurrence of events that would take place in the corresponding hardware implementation. The remainder of this section discusses the execution model for simulating circuits described in VHDL. As a practical issue, this understanding is invaluable in debugging VHDL simulations.

2.7.1   Discrete Event Simulation

The preceding sections presented a description of the behavior of digital systems in terms of events that take place at discrete points in time. Some events may cause other events to be generated after some delay, and many events may be generated concurrently. *Discrete event simulation* is a programming-based methodology for accurately modeling the generation of events in physical systems. The operation of a physical system, such as a digital circuit, is described in a computer program that specifies how and when events—changes in signal values—are generated. A *discrete event simulator* then executes this program, modeling the passage of time and the occurrence of events at various points in time. Such simulators often manage millions of events and rely on well-developed techniques to accurately keep track of the *correct order* in which the events would have occurred in the physical system. We can view VHDL as a programming language for describing the generation of events in digital systems supported by a discrete event simulator. The next section describes a simple discrete event simulation model that captures the basic elements of the simulation of VHDL programs. Understanding the VHDL model of time is a necessary prelude to writing, debugging, and understanding VHDL models.

2.7.2   A Discrete Event Simulation Model for VHDL

Discrete event simulations utilize an event list data structure. The event list is an ordered list of all future events in the circuit. Each event is described by the type of event—for example, a $0 \rightarrow 1$ or $1 \rightarrow 0$ transition—and the time at which it is to occur. Although we generally think of transitions as a change of value between 0 and 1, recall that signals may have other values. In general, an event is simply a change in the value of a signal. For example, the transition $U \rightarrow 1$ represents an event. Let us refer to the time at which an event is to occur as the *timestamp* of that event. The event list is ordered according to increasing timestamp value. This enables the simulator to execute events in the order that they occur in the real world—that is, the physical system. Note that many events may have the same timestamp value. While these events will be executed sequentially by the simulator, they will be recorded as having occurred at the same time. Timestamps correspond to the simulator clock and record the passage of simulated time. The value of this clock will be referred to as the current *timestep*, or simply timestep. Imagine what would happen if we froze the system at a timestep and took a snapshot of the values of all of the signals in the system. These values would represent the *state* of the simulation at that point in time.

Example: Discrete Event Simulation

Let us consider one approach to the discrete event simulation of the half-adder circuit shown in Figure 2.2. Assume that we have been able to specify waveforms on the inputs a and b. In a physical circuit, these waveforms would probably be generated as the output of another component. At timestep 0 ns, initial values on inputs a and b will cause events that set the values of the sum and carry signals. Assuming that the propagation delay of a gate is 5 ns, these events will be scheduled to occur 5 ns later. Figure 2.9(a) shows these events at the head of the event list at time 5 ns. Note that both sum and carry are scheduled to receive values at this time. Prior to this time, the values of the sum and carry are undefined, as represented by the shaded areas in the timing diagram shown in Figure 2.2. Input a is also scheduled to make a transition at the same time. For the moment, let us ignore how these events on the input signal are generated. These events correspond to the signal transitions shown on the timing diagram in Figure 2.2 at 5 ns. The simulator removes these events from the event list, and the current values of these signals are updated. Due to a change in the value of signal a, the simulator determines that the values of the output signals, sum and carry, have to be recomputed. The computation produces new values of sum and carry, which are scheduled in the event list at timestep 10 ns, since the gate delay is 5 ns. The head of the event list now appears as shown in Figure 2.9(b), with all of the events scheduled for timestep 10 ns. The global clock is now updated to 10 ns, all events scheduled at timestep 10 ns are removed from the event list, the corresponding signal values are updated, and any new events are computed and scheduled by insertion into the list in timestamp order. Figure 2.9(c) shows the head of the event list after event computations at timestep 10 ns and prior to

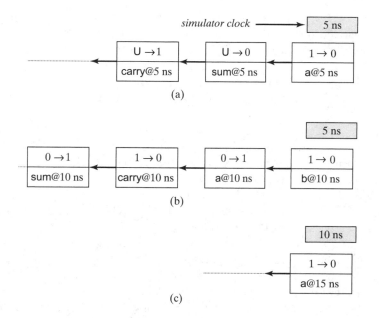

FIGURE 2.9   Discrete event simulation of the half adder

the update of the global clock. The simulator clock will now be updated to 15 ns, and the process repeats until there are no more events to be computed or until some predetermined simulation time has expired.

Example End: Discrete Event Simulation

This behavior of a discrete event simulator as captured in the preceding example can be described in the following steps:

1. Advance simulation time to that of the event with the smallest timestamp in the event list. This is the event at the head of the sorted event list.
2. Execute all events at this timestep by updating signal values.
3. Execute the simulation models of all components affected by the new signal values.
4. Schedule any future events (i.e., update of signal values) that occur as a result of step 3 by inserting events into the list in timestamp order.
5. Repeat steps 1–4 until the event list is empty or a preset simulation time has expired.

In general, there is substantial concurrency in a digital circuit and many events may take place simultaneously. Thus, many signals may receive values at the same timestep and more than one event is executed at a timestep. We see that the simulator employs a two-stage model of the evolution of time. In the first stage, simulation time is advanced to that of the next event and all signals receiving values at this time are updated. In the second stage, all components affected by these signal updates are reevaluated, and any future events that are generated by these evaluations are scheduled by placing them into the event list in order of their timestamp.

It is apparent that this model is quite flexible and general. We can think of modeling gate-level circuits as well as higher level circuits, such as arithmetic logic units (ALUs), decoders, multiplexors, and even microprocessors. We simply need to describe the behavior of these components in terms of input events, computation of the output events from input events, and propagation delays. If we can describe a digital system in these terms, we can develop computer programs to implement this behavior. Note that discrete event simulation itself is only a model or approximation of the behavior of real systems. In real circuits, signals do not make instantaneous transitions between logic 0 voltage levels and logic 1 voltage levels. For that matter, there are no such things as truly digital devices. There are only analog devices wherein we interpret analog voltage levels as 0 or 1! In practice, for many purposes, such a discrete event model is adequate. However, often more detailed and accurate models are required, in which case complementary techniques and models are employed. To distinguish the discrete event model from the real system, we will refer to the former as the *logical model* and the latter as the *physical system*.

The basic data structures and concepts are common across discrete event simulation systems. The VHDL simulators will provide facilities for setting the duration of a simulation timestep and to query the contents of the event queue at points during the simulation. The user can also typically examine the event at the head of the event queue and force signals to specific values prior to the next time step. An understanding of this underlying simulation model and the basic data structures is invaluable in debugging VHDL simulations.

2.7.3   Accuracy vs. Simulation Speed

In the context of the discrete event simulation model, it is now clear why there is a natural trade-off between simulation accuracy and simulation speed. Suppose we wish to construct a simulation of a 32-bit adder. Our first VHDL model may be at the level of an adder component that is described by the delay in computing a sum of its input values. The behavior of the box is simply to add the two inputs and produce an output value. In later chapters, we will see how we can write such a simple, one-line, VHDL model. When the two input values are available, an output value is computed and scheduled for some time, say 5 ns, into the future. The presence of input values causes one event, namely, the change in the value of the output, to be scheduled for some point in time in the future. An alternative and relatively more accurate model of the adder might describe a gate-level implementation of a 32-bit adder. Now, when new inputs become available, a sequence of events is generated along the internal signals that connect the inputs and outputs of gates in the design. We can imagine that, if we add up the total number of events that are generated in this discrete event model, it would be on the order of tens of events. Each of these events must be inserted in the event queue and eventually removed as simulated time progresses. The amount of work the simulator performs and therefore the time it spends is proportional to the number of events that are handled. One step in the simulation now will correspond to a gate delay rather than the full delay through the adder.

The key observation is that, in general, in gate-level models, the discrete event simulation may process several orders of magnitude more events than in higher level models, such as those that register transfer-level models of a hardware design. Accuracy is improved, since we monitor gate-level events. However, the amount of work that the simulator must perform is much greater, and hence simulation time increases. Scale this simple example up to the context of 10 million gate designs, and we have nonnegligible differences in simulation times!

Accuracy has traditionally been a function of simulation time, and it is not unusual for simulations to run overnight or even for days. In fact, for many years, there were vendors that sold special-purpose hardware just for logic simulation. That technology has now evolved to use new classes of commercial hardware, while many vendors still rely on the clusters of high-end workstations.

2.8   Chapter Summary

This chapter is based on the premise that, to use the VHDL language most effectively, we must understand the distinguishing physical, structural, and behavioral attributes of the digital systems that the language is designed to describe. I hope that this approach will provide an intuitive basis for the reader to learn and apply language constructs described in succeeding chapters. Further, to be able to write and debug VHDL programs, one must understand how such simulations are executed. The discrete event model described here sets the expectations of the developer and provides guidance in the form of "what are the right questions to ask" in debugging designs.

The key attributes discussed in this chapter include the following:

- System descriptions
  - interface
  - function
    - structural
    - behavioral
- Events
- Propagation delays
- Concurrency
- Timing
  - synchronous
  - asynchronous

- Waveforms
- Signal values
- Shared signals
- Discrete event simulation
  - events and timestamps
  - event list and timestep
  - simulation speed vs. accuracy

The VHDL language provides basic constructs for representing each of the preceding attributes. An associated simulator implements a discrete event simulation model, manages the progression of simulated time, and maintains internal representations of the waveforms being generated on signals. Synthesis compilers process the VHDL descriptions and a set of target hardware components to create a circuit that implements the behavior captured in the VHDL description. The next chapter describes these underlying simulation and synthesis models, while the remaining chapters present language constructs to specify these attributes.

CHAPTER 3

Basic Language Concepts

This chapter introduces the basic language constructs of the VHDL language. Our goal is to construct models of digital systems for the purpose of simulating their behavior. Our mind-set is that of *describing* a physical design rather than a computation. An associated simulator is concerned with *how* the design is simulated.

VHDL has often been criticized as being overly complex and intimidating to the novice user. While the language is extensive, a quick start towards building useful simulation models can be made by relying on a core set of language constructs. This chapter discusses the language constructs provided within VHDL for describing the attributes of digital systems identified in Chapter 2, such as events, propagation delays, concurrency, and waveforms. Chapter 4 then introduces concepts that extend the constructs introduced here so that they enable us to apply conventional programming constructs in building models of complex digital systems, particularly at higher levels of abstraction. Collectively, these two chapters provide us with the tools necessary to model all of the attributes of digital systems described in Chapter 2.

3.1   Signals

Fundamentally, digital systems are about signals—transporting and operating on signals. We might then expect that the notion of a signal is a basic part of any language for describing digital hardware. Conventional programming languages manipulate basic objects such as variables and constants. Variables receive values through assignment statements and can be assigned new values through the course of a computation. Constants, on the other hand, may not change their values. To capture the behavior of digital signals, the VHDL language introduces a new type of programming object: the signal object type.

We have seen that signals may take on one of several values, such as 0, 1, or Z. Signals are analogous to the wires used to connect components of a digital circuit. Like variables, signals may also be assigned values, but they differ from variables in that they have an associated *time value,* since a signal receives a value at a specific point in time. The signal retains this value until it is assigned a new value at a future point in time. The sequence of values assigned to a signal over time is the *waveform* of the signal. It is primarily this association with time–value pairs that differentiates a signal from a variable. A variable always has one current value. At any instant in time a signal may be associated with several time–value pairs, where each time–value pair represents some future value of the signal. (Remember, we are simulating the behavior of the signal over time.) Finally, note that a variable may be declared to be of a specific type, such as **integer**, **real**, or **character**. In a similar manner, a signal can be declared to be of a specific type. When used in this way, a signal does not necessarily correspond to wires that connect digital components. For example, we may model the output of an arithmetic logic unit (ALU) as an integer-valued signal. This output is treated as a signal and, in simulation, behaves as a signal by receiving values at specific points in time. However, we do not have to concern ourselves with modeling the number of bits necessary at the output of the ALU as we would if we were modeling the ALU at a much lower level, for example, at the gate level. When we are building simulation models of components, we often are not interested in the implementation of the ALU as much as we are interested in capturing this behavior accurately in a VHDL model. Thus, we are not required to think of signals in terms of a number of bits. This behavior enables us to model systems at a higher level of abstraction than digital circuits. Such high-level simulation is useful in the early stages of the design process, where many details of the design are still being developed.

Before we can understand how to declare and operate on signals, we must cover the basic programming constructs in VHDL. We will return to discuss signal objects in greater detail later in this chapter.

3.2   Entity–Architecture

We start by addressing the issue of describing digital systems. The primary programming abstraction in VHDL is a *design entity*. Examples of design entities include a chip, board, and transistor. A design entity is a component of a design whose behavior is to be described and simulated.   Consider once again the gate-level digital circuit for a half adder shown in Figure 3.1. There are two input signals, x and y. The circuit computes the value of two output signals, sum and carry. This half-adder circuit represents an example of a design entity.

How can the half-adder circuit be accurately described? Imagine that you had to describe this circuit over the telephone to a friend who was familiar with digital logic gates, but was not familiar with the half adder. Your description would most likely include the input signals, the output signals, and a description of the behavior. The behavior in turn may be specified with a truth table, Boolean equations, or simply an interconnection between gates. We observe that there are two basic components to the

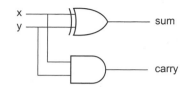

FIGURE 3.1 Half-adder circuit

description of any design entity: (i) the interface to the design and (ii) the internal behavior of the design. The VHDL language provides two distinct constructs to specify the interface and internal behavior of design entities, respectively.

The external interface to this entity is specified with the **entity** declaration. For the circuit shown in Figure 3.1, the entity declaration would appear as follows:

entity half_adder **is**
port(x, y: **in bit**;
        sum, carry: **out bit**);
end entity half_adder;

The boldface type denotes keywords that are VHDL reserved keywords. The remaining are user supplied. Just as we name programs, the label half_adder is the name given to this design entity by the programmer. Names can consist of upper- or lowercase characters or digits and may include the underscore character "_" as in the preceding code. However, hyphens ("-") are not permitted, the first character of a user-supplied name must be a letter, and the last character cannot be an underscore. The VHDL language is *case insensitive*. Thus, half_adder and HALF_ADDER would refer to the same entity.

The inputs and outputs of the circuit are referred to as *ports*. The ports are special programming objects and are signals. Ports are the means by which the half adder can communicate with the external world and other circuits. Therefore, naturally, we expect ports to be signals rather than variables. Like variables in conventional programming languages, each port must be a signal that is declared to be of a specific type. In this case, each port is declared to be of type **bit** and represents a single-bit signal. A **bit** is a signal type defined within the VHDL language and can take the values of 0 or 1. A **bit_vector** is a signal type composed of a vector of signals, each of type **bit**. The type **bit** and **bit_vector** are two common types of ports. In general, a port may be any one of several other VHDL data types. Chapter 9 describes common data types and operators supported by the language.

The original VHDL language standard is referenced as the 1987 standard 1076–1987. Modifications to the language were ratified into the 1993 standard, referred to as 1076–1993. This text follows the 1993 standard. Most differences introduced in VHDL'93 deal with introduction of a few new concepts, tuning the semantics of existing constructs, and refining the syntax of the major constructs. However, in the interests of compatibility with legacy code, we will identify differences between the VHDL'87 and VHDL'93. For example, there is a difference in the syntax of the **entity** construct.

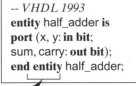

```
-- VHDL 1993
entity half_adder is
port (x, y: in bit;
sum, carry: out bit);
end entity half_adder;
```

new

```
-- VHDL 1987
entity half_adder is
part (x, y: in bit;
sum, carry: out bit);
end half_adder;
```

From our study of digital logic, we know that bits and bit vectors are fundamental signals. From Chapter 2, we know that, from the perspective of simulation, we are interested in many more values of signals. For example, a signal may be uninitialized or not driven to a voltage level denoting logic 0 or logic 1 and thus will be in a high impedance state denoted by Z. Hence, in practice, the types **bit** and **bit_vector** are of limited use. How many values should a signal have? The problem was that vendors began defining new signal types. For example, vendor SimVHDL Inc. defines a new type called RealSignal. All signals in a VHDL model using this company's simulator are defined to be of type RealSignal, and such signals can take on, say, one of 12 values. None of the models written for this simulator can be used with any other simulator, unless they support the type RealSignal, which in all likelihood they will not, since it is not defined as part of the language. This eliminates one big hope for VHDL, namely, reuse of models across vendors.

In steps the IEEE again and coordinates the development of a standard type system and associated supporting implementations. The IEEE 1164 Standard has gained widespread acceptance as a standard value system. This standard defines a nine-value signal as shown in Figure 2.6. To use this value system, signals would be declared to be of type std_ulogic rather than **bit**. Analogously, we would have the type std_ulogic_vector rather than **bit_vector**. Therefore, throughout the remainder of this text, all examples will utilize the IEEE 1164 standard signal and data types. The preceding entity declaration would now appear as follows:

```
entity half_adder is
port (x, y: in std_ulogic;
sum, carry: out std_ulogic);
end entity half_adder;
```

The signals appearing in a port declaration may be distinguished as input signals, output signals, or bidirectional signals. This is referred to as the *mode* of the signal. In the preceding example, the **in** and **out** specifications denote the mode of the signal. Bidirectional signals are of mode **inout**. Every port in the entity description must have its mode and type specified.

We see that it is relatively straightforward to write the entity descriptions of standard digital logic components. We next show some sample circuits and their entity descriptions. Note how byte and word-wide groups of bits are specified. For example, a

32-bit quantity is declared to be of the type std_ulogic_vector (31 **downto** 0). This type refers to a data item that is 32 bits long, where bit 31 is the most significant bit in the word and bit 0 is the least significant bit in the word.

Example: Entity Declaration of a 4-to-1 Multiplexor

  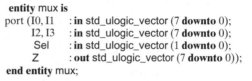

```
entity mux is
  port (I0, I1    : in std_ulogic_vector (7 downto 0);
        I2, I3    : in std_ulogic_vector (7 downto 0);
        Sel       : in std_ulogic_vector (1 downto 0);
        Z         : out std_ulogic_vector (7 downto 0));
  end entity mux;
```

Example End: Entity Declaration of a 4-to-1 Multiplexor

Example: Entity Declaration of a D Flip-Flop

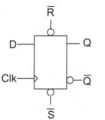

```
entity D_ff is
  port (D, Q, Clk, R, S : in std_ulogic;
        Q, Qbar         : out std_ulogic);
  end entity D_ff;
```

Example End: Entity Declaration of a D Flip-Flop

Example: Entity Declaration of a 32-Bit ALU

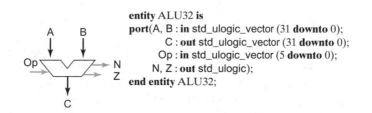

```
entity ALU32 is
  port(A, B : in std_ulogic_vector (31 downto 0);
       C : out std_ulogic_vector (31 downto 0);
       Op : in std_ulogic_vector (5 downto 0);
       N, Z : out std_ulogic);
  end entity ALU32;
```

Example End: Entity Declaration of a 32-Bit ALU

From the preceding examples, it is clear that design entities can occur at multiple levels of abstraction, from the gate level to large systems. In fact, it should be apparent that a design entity does not even have to represent digital hardware! The description of the interface is simply a specification of the input and output signals of the design entity.

Once the interface to the digital component or circuit has been described, it is necessary to describe its internal behavior. The VHDL construct that enables us to specify the behavior of a design entity is the **architecture** construct. The syntax of the architecture construct is the following:

```
-- VHDL 1993

architecture behavioral of half_adder is
-- place declarations here
begin
-- place description of behavior here --
end architecture behavioral;
```

new

```
-- VHDL 1987

architecture behavioral of half_adder is
-- place declarations here
begin
-- place description of behavior here --
end behavioral;
```

'87 vs. '93

The preceding construct provides for the declaration of the module named behavioral, which will contain the description of the behavior of the design entity named half_adder. Such a module is referred to as the **architecture** and is associated with the entity named in the declaration. Thus, the description of a design entity takes the form of an entity–architecture pair. The architecture description is linked to the correct entity description by providing the name of the corresponding entity in the first line of the architecture. The same rules for constructing entity names apply to architecture names.

The behavioral description provided in the architecture can take many forms. These forms differ in the levels of detail, description of events, and the degree of concurrency. The remainder of this chapter focuses on a core set of language constructs required to model the attributes of digital systems described in Chapter 2. Subsequent chapters will add constructs, motivated by the need for expanding the scope and level of abstraction of the systems to be modeled.

3.3   Concurrent Statements

The operation of digital systems is inherently concurrent. Many components of a circuit can be operating simultaneously and concurrently driving distinct signals to new values. How can we describe the assignment of values to signals? We know that signal values are time–value pairs; that is, a signal is assigned a value at a specific point in time. Within VHDL, signals are assigned values using *signal assignment* statements. These statements specify a new value of a signal and the time at which the signal is to acquire this value. Multiple signal assignment statements are executed concurrently in simulated time and are referred to as *concurrent signal assignment statements* (*CSA*s). There are several forms of **CSA** statements, described next.

3.3.1 Simple Concurrent Signal Assignment

Consider a description of the behavior of the half-adder circuit shown in Figure 3.1. Recall that although VHDL manages the progression of time, we need to be able to specify events, delays, and concurrency of operation.

```
architecture concurrent_behavior of half_adder is
begin
 sum <= (x xor y) after 5 ns;
 carry <= (x and y) after 5 ns;
end architecture concurrent_behavior;
```

Just as we named entity descriptions, the label concurrent_behavior is the name given to this architecture module. The first line denotes the name of the entity that contains the description of the interface for this design entity, which in this case is the entity half_adder. Each statement in the preceding architecture is a *signal assignment* statement with the operator "<=" denoting signal assignment. Each statement describes how the value of the output signal depends on, and is computed from, the value of the input signals. For example, the value of the sum output signal is computed as the Boolean exclusive-OR operation of the two input signals. Once the value of sum has been computed, it will not change unless the value of x or y changes. Figure 3.2 illustrates this behavior. At the current time, x = 0, y = 1, and sum = 1. At time 10, the value of y changes to 0. The new value of the sum will be (x **xor** y) = 0. Since there will be a propagation delay through the exclusive-OR gate, the signal sum will be assigned this value 5 ns later, at time 15. This behavior is captured in the first signal assignment statement. Note that, unlike variable assignment statements, the signal assignments shown specify both value and (relative) time.

In general, if an event (signal transition) occurs on a signal on the right-hand side of a signal assignment statement, the expression is evaluated and new values for the output signal are scheduled for some time in the future as defined by the **after** keyword. The dependency of the output signals on the input signals is captured in the two statements and **NOT** in the textual order of the program. The textual order of the statements could be reversed and the behavior of the circuit would not change. Both statements are executed concurrently with respect to simulated time to reflect the concurrency of the corresponding operations in the physical system. This is why these statements are referred to as concurrent signal assignment statements. A fundamental difference between VHDL programs and conventional programming languages is that concurrency is a natural part of the systems described in VHDL and therefore of the language itself. Note that the flow of signal values, rather than textual order, determines the execution of the statements. Figure 3.2

```
library IEEE;
use IEEE.std_logic_1164.all;

entity full_adder is
port (x, y: in std_ulogic;
      sum, carry : out std_ulogic);
end entity half_adder;

architecture concurrent_behavior of half_adder is
begin
sum <= (x xor y) after 5 ns;
carry <= (x and y) after 5 ns;
end architecture concurrent_behavior;
```

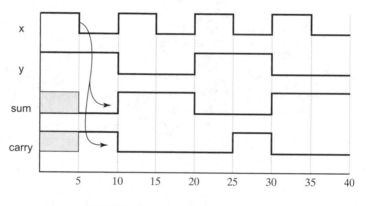

FIGURE 3.2   Operation of a half adder

shows a complete, executable half-adder description and the associated timing behavior. This description contains the most common elements used to describe a design entity.

Note the use of the **library** and **use** clauses. We can think of libraries as repositories for frequently used design entities that we wish to share. The **library** clause identifies a library that we wish to access. The name is a logical name for a library. In Figure 3.2, the library name is IEEE. In practice, this logical name will usually map to a directory on the local system. This directory will contain various design units that have been compiled. A *package* is one such design unit. A package may contain definitions of types, functions, or procedures to be shared by multiple application developers. The **use** clause determines which of the packages or other design units in a library you will be using in the current design. In the preceding example, the use clause states that in library IEEE, there is a package named std_logic_1164 and that we will be able to use all of the components defined in this package. We need the package, since the definition for the type std_ulogic is in it and is not a part of the language definition. The VHDL models that use the IEEE 1164 value system will include the package declaration as shown. Design tool vendors typically provide the library **IEEE** and the **std_logic_1164** package. These concepts are analogous to the use of libraries for mathematical functions and input–output in conventional programming languages. Libraries and packages are described in greater detail in Chapter 6. This example now contains the major components found in VHDL models: declarations of existing design units in libraries that you will be using, the entity description of the design unit, and the architecture description of the design unit.

The descriptions provided so far in this chapter are based on the specification of the value of the output signals as a function of the input signals. In larger and more complex designs, there are usually many internal signals used to connect design components, such as gates or other hardware building blocks. The values that these signals acquire can also be written using simple concurrent signal assignment statements. However, we must be able to declare and make use of signals other than those within the entity description. The gate-level description of the full adder provides an example of such a VHDL model.

Example: Full-Adder Model

Consider the full-adder circuit shown in Figure 3.3. We are interested in an accurate simulation of this circuit in which all of the signal transitions in the gate-level realization are modeled. In addition to the ports in the entity description, we see that there are three internal signals. These signals are named and declared in the architectural description. The declarative region declares three single-bit signals: s1, s2, and s3. These signals are annotated in the circuit. Now we are ready to describe the behavior of the full adder in terms of the internal signals as well as the entity ports. Since this circuit uses two input gates, each signal is computed as a Boolean function of two other signals. The model is a simple statement of *how* each signal is computed as a function of other signals, and the propagation delay through the gate. There are two output signals and three internal signals, for a total of five signals. Accordingly, the description consists of five concurrent signal assignment statements, one for each signal.

Each signal assignment statement is given a label: L1, L2, and so on. This labeling is optional and can be used for reference purposes. Note a new language feature in this model: the use of the **constant** object. Constants in VHDL function in a manner similar to the way they do in conventional programming languages. A constant can be declared to be of a specific type, in this case of type **Time**. A constant must have a value at the start of the simulation and cannot be changed during the simulation. At this stage, it is easiest to ensure that constants are initialized as shown in Figure 3.3. The introduction of the type **Time** is a natural consequence of simulation modeling. Any object of this type must take on the values of time, such as microseconds or nanoseconds. The type **Time** is a predefined type of the language. As we know, the textual order of the statements is irrelevant to correct operation of the circuit model. Let us now consider the flow of signal values and the sequence of execution of the signal assignment statements. Figure 3.4 shows the waveforms of all of the signals in the full-adder circuit. From the

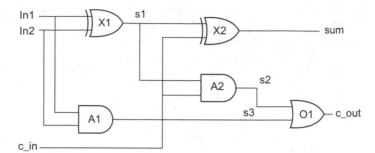

```
library IEEE;
use IEEE.std_logic_1164.all;
entity full_adder is
port (in1, in2, c_in: in std_ulogic;
     sum, c_out: out std_ulogic);
end entity full_adder;

architecture dataflow of full_adder is
signal s1, s2, s3 : std_ulogic;                  Architecture
constant gate_delay: Time:= 5 ns;                Declarative
begin                                            Segment
L1: s1 <= (In1 xor In2) after gate_delay;
L2: s2 <= (c_in and s1) after gate_delay;        Architecture
L3: s3 <= (In1 and In2) after gate_delay;        Body
L4:  sum <= (s1 xor c_in) after gate_delay;
L5:  c_out <= (s2 or s3) after gate_delay;
end architecture dataflow;
```

FIGURE 3.3   VHDL model of a full adder

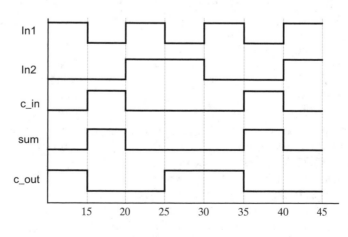

FIGURE 3.4   Full-adder circuit timing

figure, we see that there is an event on In1 at time 10 changing its value to 1. This causes statements L1 and L3 (from Figure 3.3) to be executed and new values to be scheduled on signals s1 and s3 5 ns later, at time 15 ns. These events in turn cause statements L2 and L5 to be executed at time 20 ns and events to be scheduled on signals c_out and s2 at time 20 ns. We see that the execution of the statement L1 produced events that caused the execution of statement L5. This order of execution is maintained regardless of the textual order in which the statements appear in the program.

Note the two-stage model of time. In the first stage, all statements with events occurring at the current time on signals on the right-hand side (RHS) of the signal assignment statement are evaluated. All future events that are generated from the execution of these statements are then scheduled. Time is now advanced to the time of the next event. The process repeats. Note how the programmer specifies events, delays, and concurrency. He or she specifies events with signal assignment statements, delays within the signal assignment statement relative to the current time, and concurrency by a distinct signal assignment statement for each signal. The order of execution of the statements is dependent upon the flow of values (just as is the case in the real circuit) and not on the textual order of the program. As long as the programmer correctly specifies how the value of each signal is computed and when it acquires this value relative to the current time, the simulator will correctly reflect the behavior of the entire circuit.

Example End: Full-Adder Model

3.3.2 Implementation of Signals

Unlike variables, signals are a new type of programming object and merit specific attention. For example, we know how variables are implemented: They are simply a location in memory. If a variable is assigned a value, the corresponding location in memory is written with the new value while destroying the old value. This effectively happens immediately so that, if the next executing statement in the program uses the value of the variable, it is the new value that is used. A signal is a different type of object. We naturally think of signals as having a history of values over time—for example, as a waveform. If we are to preserve this intuition, then the internal storage mechanism for signals must be quite different from that employed for variables.

So far, we have seen that signals can be declared in the body of an architecture or in the port declaration of an entity. The form of the declaration is

signal s1 : std_ulogic := '0';

If the signal declaration includes the assignment symbol (i.e., :=) followed by an expression, the value of the expression is the initial value of the signal. The initialization is not required, in which case the signal is assigned a default value as specified by the type definition. For example, all Boolean valued signals may be assigned FALSE initially. Signals can be declared to be one of many valid VHDL types: **integers**, **real**, **bit_vector**, and so forth.

Now consider the assignment of values to a signal. We know that signal assignment statements assign a value to a signal at a specific point in time. The simple concurrent signal assignment statements described so far in this chapter exhibit the following common structure:

sum <= (x **xor** y) **after** 5 **ns**;

which can be written in a more general form as

signal <= *value expression* **after** *time expression*;

The expression on the right-hand side of the signal assignment is referred to as a *waveform element*. A waveform element describes an assignment to a signal and is composed of a *value expression* to the left of the **after** keyword and a *time expression* to the right of the keyword. The former evaluates to the new value to be assigned to the signal and the latter evaluates to the relative time at which the signal is to acquire this value. In this case, the new value is computed as the exclusive-OR of the current values of the signals x and y. The value of the time expression is added to the current simulation time to determine when the signal will receive this new value. In this case, the time expression is a constant value of 5 ns. With respect to the current simulation time, this time–value pair represents the future value of the signal and is referred to as a *transaction*. The underlying discrete event simulator that executes VHDL programs must keep track of all transactions that occur on a signal. The list is ordered in increasing time of the transactions.

If the evaluation of a single waveform element produces a single transaction on a signal, can we specify multiple waveform elements and, as a result, multiple transactions? For example, could we have the following?

s1 <= (x **xor** y) **after** 5 **ns**, (x **or** y) **after** 10 **ns**, (**not** x) **after** 15 **ns**;

The answer is, yes! When an event occurs on either of the signals x or y, the preceding statement will be executed, all three waveform elements will be evaluated, and three transactions will be generated. Note that these transactions are in increasing order of time. The events represented by these transactions must be scheduled at different points in the future, and the VHDL simulator must keep track of all of the transactions that are currently scheduled on a signal. The simulator does that by maintaining an ordered list of all of the current transactions pending on a signal. This list is referred to as the *driver* for the signal. The current value of a signal is the value of the transaction at the head of the list. What is the physical interpretation of such a sequence of events? These events represent the value of the signal over time, which essentially is a waveform. This is how we can represent a signal waveform in VHDL: as a sequence of waveform elements. Therefore, within a signal assignment statement, rather than assigning a single value to the signal at some future time, we can assign a waveform to this signal. We specify the waveform as a sequence of signal values, and we specify each value with a single waveform element. Within the simulator, these sequences of waveform elements are represented as a sequence of transactions on the driver of the signal. The transactions are referred to as the *projected output waveform*, since these events have not yet occurred in the simulation. What if the simulation attempts to add transactions that conflict with the current projected waveform? The VHDL language definition provides specific rules for adding transactions to the projected waveform of a signal. For a precise definition of the rules the reader can refer to the language reference manual, or LRM [8].

Example: Specifying Waveforms

Assume that we would like to generate the following waveform:

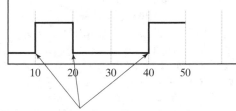

signal transitions for each waveform element

We could do so with the following signal assignment:

> signal <= '0','1' **after** 10 **ns**,'0' **after** 20 **ns**,'1' **after** 40 **ns**;

Note how each transition in the preceding waveform is specified as a single waveform element in the signal assignment statement. All waveform elements must be ordered in increasing time. Failure to do so will result in an error.

Example End: Specifying Waveforms

The concepts and terminology discussed so far are derived from the operation of digital circuits. In a physical circuit, a wire (signal) has a driver associated with it. Over time, this driver produces a waveform on that wire. If we continue to view the language constructs by analogy with the digital circuits they were intended to model, it will be easier for us to reason about the construction of models using VHDL. The constructs that manipulate signals invariably rely on waveform elements to specify input and output waveforms. Understanding this representation is key to understanding many of the VHDL programming constructs.

3.3.3   Resolved Signals

Our view of signals up to this point has been one in which every signal has only one driver—that is, one signal assignment statement that is responsible for generating the waveform on that signal. We know that is not true in practice. Shared signals occur on buses and in circuits based on wired logic. When a signal has multiple drivers, how is the value of the signal determined? In the VHDL language, this value is determined by a *resolution function*.

A resolution function examines all of the drivers on a shared signal and determines the value to be assigned to the signal. A shared signal must be of a special type: a *resolved type*. A resolved type has a resolution function associated with the type. In the preceding examples, we have been using the std_ulogic and std_ulogic_vector types for single-bit and multibit signals, respectively. The corresponding resolved types are std_logic and std_logic_vector. This distinction has a number of consequences. In the course of the simulation, when a signal of type std_logic is assigned a value, the associated resolution function is automatically invoked to determine the correct value of the signal. Multiple drivers for this signal may be projecting multiple future values for this signal. The resolution function examines these drivers to return the correct value of the signal at the current time. If the signal has only one driver, then the value is determined straightforwardly. However, if more than one driver exists for the signal, the value that is assigned to the signal is the value determined by the resolution function. For the IEEE 1164 package, the resolution function is essentially a lookup table. Provided with the signal values from two drivers, the table returns the signal value to be assigned. For example, if one source is driving the signal to 1 and a second source's output is left floating (i.e., in state Z), the resulting value will be 1. Alternatively, if the two sources are driving the shared signal to 1 and 0, respectively, the resulting value will be unknown or X. The resolution function for the std_logic and std_logic_vector types is provided by the std_logic_1164 package. Having multiple drivers for a signal whose type is an unresolved type will result in an error. The user may define new resolved types and provide the resolution functions for their use.

We will leave resolution functions for the moment and return to them in greater detail when we deal with the creation and use of procedures and functions in Chapter 6. However, in the remainder of this text, all of the examples use the IEEE 1164 resolved single-bit and multibit types, std_logic and std_logic_vector, rather than unresolved types std_ulogic or std_ulogic_vector.

Simulation Exercise 3.1: A First Simulation Model

This exercise introduces the construction and simulation of simple VHDL models. The following steps require simulator-specific commands:

Step 1.   Using a text editor, create a VHDL model of the full adder shown in Figure 3.3. Do not use a word processor, even though it may have an option for saving your text as an ASCII file. Some word processors place control characters in the file or may handle some characters nonuniformly—for example, left and right quotation marks. This can lead to analyzer errors. (However, you could correct these errors at that time.) Set the gate delays to 3 ns for the EX-OR gates and to 2 ns for all of the other gates.

Step 1(a) Declare and reference the library IEEE and the package std_logic_1164.

Step 1(b) Write the entity description. Use the types std_logic and std_logic_vector for the input and output signals.

Step 1(c) Write the architecture description.

Step 2. Compile and load the model for simulation using a VHDL simulator toolset.

Step 3. Generate a waveform on each of the input signals.

Step 4. Run the simulation for 40 ns and trace (i) the input signals, (ii) the internal signals, s1, s2, and s3, and (iii) the sum and carry outputs.

Step 5. Check and list scheduled events on the internal signals and output signals.

Step 6. Pick an event on one of the input signals. Record the propagation of the effect of this event through the signal trace. Study the trace and ensure that the model is operating correctly.

Step 7. Repeat this example, only this time do not initialize one of the input signals. What does the resulting trace look like and what is the significance of the values on this uninitialized input?

End Simulation Exercise 3.1

3.3.4 Conditional Signal Assignment

The simple concurrent signal assignment statements that we have seen so far compute the value of the target signal on the basis of Boolean expressions. The values of the signals on the RHS of the signal assignment statement are used to compute the value of the target signal. This new value is scheduled at some point in the future, using the **after** keyword. Expressing values of signals in this manner is convenient for describing gate-level circuits whose behavior can be expressed with Boolean equations. However, we often find it useful to model circuits at higher levels of abstraction, such as multiplexors and decoders. Modeling at this level requires a richer set of constructs.

For example, consider the physical behavior of a 4-to-1, 8-bit multiplexor, as shown in Figure 3.5. The value of Z is one of In0, In1, In2, or In3. The waveform that appears on one of the inputs is transferred to the output Z. The specific choice depends upon the value of the control signals S0 and S1, for which there are four possible alternatives. Each of these must be tested and one chosen. This behavior is captured in the conditional signal assignment statement and is illustrated for a 4-to-1, 8-bit multiplexor in Figure 3.5. The structure of the statement follows from the physical behavior of the circuit. For each of the four possible values of S0 and S1, an input waveform is specified. In this case, the waveform consists of a single waveform element describing

```
library IEEE;
use IEEE.std_logic_1164.all;
entity mux4 is
port (In0, In1, In2, In3 : in std_logic_vector (7 downto 0);
      S0, S1: in std_logic;
      Z : out std_logic_vector (7 downto 0));
end entity mux4;

architecture behavioral of mux4 is
begin
Z <= In0 after 5 ns when S0 = '0' and S1 = '0' else
     In1 after 5 ns when S0 = '0' and S1 = '1' else
     In2 after 5 ns when S0 = '1' and S1 = '0' else
     In3 after 5 ns when S0 = '1' and S1 = '1' else
     "00000000" after 5 ns;
end architecture behavioral;
```

FIGURE 3.5   Conditional signal assignment statement

the most recent signal value on that input. As pointed out in Section 3.3.2, more than one waveform element in each line of the conditional statement could have been specified, producing a waveform on the output signal Z.

In the corresponding physical circuit, an event on any one of the input signals, In0–In3, or any of the control signals, S0 or S1, may cause a change in the value of the output signal. Therefore, whenever any such event takes place, the concurrent signal assignment statement is executed and all four conditions may be checked. The order of the statements is important. The expressions in the RHS are evaluated in the order that they appear. The first conditional expression that is found to be true determines the value that is transferred to the output. Therefore, we must be careful in ordering the conditional expressions on the RHS to reflect the order in which they would be evaluated in the corresponding physical system. A careful look at the example in Figure 3.5 will reveal that, in this case, only one expression can be true and therefore the order does not matter in this particular example. Finally, note that in Figure 3.5, even though there are several lines of text, this corresponds to only one signal assignment statement.

Figure 3.6 offers a better illustration of the effect of the priority order of the expressions. The example shown represents a model of a 4-to-2 priority encoder, which produces the binary encoding of the bit that is set. If more than one of the input bits has a value of 1, then the lowest-numbered input has higher priority. In this example, if S0 is set, then the output is 00, regardless of the state of the other input bits. The last statement sets the output value to 00 and is important, since the input signals and select signals are of type std_logic. Thus, each of the select signals, S0, S1, S2, and S3, can have more values than simply 0 or 1, and the conditional signal statement shown in the example does not cover all cases. The last statement covers all of the conditions not covered by the other statements in this example.

```
library IEEE;
use IEEE.std_logic_1164.all;
entity pr_encoder is
port (S0, S1, S2, S3: in std_logic;
      Z : out std_logic_vector (1 downto 0));
end entity pr_encoder;

architecture behavioral of pr_encoder is
begin
Z <= "00" after 5 ns when S0 = '1' else
     "01" after 5 ns when S1 = '1' else
     "10" after 5 ns when S2 = '1' else
     "11" after 5 ns when S3 = '1' else
   "00" after 5 ns;
end architecture behavioral;
```

FIGURE 3.6   Priority behavior of the conditional signal assignment

Finally, the previous forms of the conditional signal assignment statement always compute a value for the output signal whenever any of the input signals change value. While this is the model for combinational logic, for more general modeling, we may require behavior wherein the output signal value may remain unchanged. This behavior can be realized with the **unaffected** keyword. For example, the conditional signal assignment statement may appear as follows:

'87 vs. '93

```
        Z <= "00" after 5 ns when S0 = '1' else
             "01" after 5 ns when S1 = '1' else
             unaffected when S2 = '1' else
             "11" after 5 ns when S3 = '1' else
             "00" after 5 ns;
```

This statement now has the following semantics: When S2 has the value 1 and both S0 and S1 have the value 0 (remember the priority ordering!), the value of the output signal does not change. This enables the description of a richer set of responses to combinations of input signal values. Such a feature—that is, the **unaffected** keyword—is not supported in VHDL'87.

3.3.5   Selected Signal Assignment Statement

The selected signal assignment statement is similar to the conditional signal assignment statement. The value of a *select expression* determines the value of a signal. For example, consider the operation of reading the value of a register from a register file with eight registers. Depending upon the value of the address, the contents of the appropriate register are selected. An example of a read-only register file with two read ports is shown in Figure 3.7.

```
library IEEE;
use IEEE.std_logic_1164.all;

entity reg_file is
port (addr1, addr2: in std_logic_vector (2 downto 0);
      reg_out_1, reg_out_2: out std_logic_vector (31 downto 0));
end entity reg_file;

architecture behavior of reg_file is
signal reg0, reg1, reg2, reg3: std_logic_vector (31 downto 0):= x"12345678";
begin
with addr1 select
reg_out_1 <= reg0 after 5 ns when "000",
                  reg1 after 5 ns when "001",
                  reg2 after 5 ns when "010",
                  reg3 after 5 ns when "011",
                  reg3 after 5 ns when others;
with addr2 (1 downto 0) select
reg_out_2 <= reg0 after 5 ns when "00",
                  reg1 after 5 ns when "01",
                  reg2 after 5 ns when "10",
                  reg3 after 5 ns when "11",
                  reg3 after 5 ns when others;
end architecture behavior;
```

FIGURE 3.7   Selected signal assignment statement

This statement operates very much like a case statement in conventional languages. As a result, its semantics is somewhat distinct from conditional signal assignment statements. For example, in Figure 3.7, the choices for the register addresses are not evaluated in sequence. Rather, all choices are evaluated, but only one must be true. Furthermore, all of the choices that the programmer specifies must cover all of the possible values of the addresses. For example, consider the VHDL code shown in Figure 3.7. Assume that we have only four registers, but both addr1 and addr2 are 3-bit addresses and therefore can address up to eight registers. The VHDL language requires you to specify the action to be taken if addr1 or addr2 takes on any of the 8 values, including those between 4 and 7. This is realized with the use of the **others** clause, as shown in the first selected signal assignment statement. The **others** keyword is used to conveniently state the value of the target signal over the remaining unspecified range of values and thereby cover the whole range. This is not really restrictive, since, in practice, we must consider what would happen in the physical system in this case. The select expression can be quite flexible and can be specified in any number of forms. For example, the select expression may incorporate Boolean expressions or, as shown in the second statement in the example, a subset of bits as in addr2(1 **downto** 0). In this latter case, do we still need the **when others** clause since addr2(1 **downto** 0) can only have 4 values? Yes, because addr2 is of type std_logic_vector. Therefore, each bit actually can take on 9 values, and addr2(1 **downto** 0) can actually have not 4, but 9^2, or 81, values!

As with simple and conditional CSAs, we must be aware of the conditions under which a selected signal assignment statement is executed. When an event occurs on a signal used in the select expression or any of the signals used in one of the choices, the statement is executed. This follows the expected behavior of the corresponding physical implementation in which an event on any of the addresses or register contents could change the value of the output signal. As with the conditional signal assignment statement, we can use the **unaffected** clause to signify that the signal does not change value. For example, the third option in the statement may be **unaffected when** "11", rather than reg3 **after** 5 **ns when** "11". Recall that this feature is supported only in VHDL 1993.

'87 vs. '93

Note a few new statements in this example. First, we initialize the values of the registers when they are declared. In the example, the registers are initialized with a hexadecimal value denoted by x"12345678". Note that the target is a signal of type std_logic_vector. In some older simulators, the hexadecimal values must be converted to the type std_logic_vector before they can be assigned. In this case, the initialization value may have to read to_stdlogicvector(x"12345678"). On the other hand, if the values were specified in binary notation explicit type conversion would not be required. The function to_stdlogicvector () is in the package std_logic_1164 and performs this type conversion operation. While this was not necessary for the simulator we used (Active VHDL and Foundation Express), that may not be the case for other simulators. Practically all CAD tool vendors support the std_logic_1164 package, which provides for type conversion as well as other functions in support of the IEEE 1164 value system. There are also other packages of functions and procedures that are provided by the vendors. Many standardization efforts are under way within the community in an effort to ensure portability of models among vendor toolsets and cooperating designers. These packages also provide similar type conversion functions, often with slightly different names. Check the availability of such packages within your toolset, and browse through them. Packages may be located in the library IEEE. If packages you wish to use are located in another design library, a new **library** clause is required to declare this library, and the **use** clause must be appropriately modified to reference this library, just as shown in our examples with IEEE. Chapter 6 describes the use of libraries and packages in greater detail.

3.4   Constructing VHDL Models Using CSAs

Armed with concurrent signal assignment statements, we are now ready to construct VHDL models of interesting classes of digital systems. This section provides a prescription for constructing such VHDL models. By following this approach to constructing models, we can generate an intuition about the structure of VHDL programs and the utility of the language constructs discussed so far.

In a VHDL model written using only concurrent signal assignment statements, the flow of data or signal values, rather than the textual order of the statements, initiates the execution of a signal assignment statement. On the basis of the language features we have seen thus far, a model of a digital system will be composed of an entity–architecture pair. The architecture model, in turn, will consist of some combination of simple, conditional, and selected signal assignment statements. The architecture may also declare and use internal signals in addition to the input and output ports declared in the entity description.

The description presented next assumes that we are writing a VHDL model of a gate-level, combinational circuit. However, the approach can certainly be applied to higher level systems using combinational building blocks such as encoders and multiplexors. The simple methodology comprises two steps: (i) the drawing of an annotated schematic and (ii) the conversion to a VHDL description. The following procedure outlines a few simple steps to organize the information we have about the physical system prior to writing the VHDL model.

Construct_Schematic

1. Represent each component (e.g., a gate) of the system to be modeled as a *delay element*. The delay element simply captures all of the delays associated with the computation represented by the component and the propagation of signals through the component. For each output signal of a component, associate a specific value of delay through the component for that output signal.
2. Draw a schematic interconnecting all of the components. Label each component uniquely.
3. Identify the input signals of the circuit as input ports.
4. Identify the output signals of the circuit as output ports.
5. All remaining signals are internal signals.
6. Associate a type with each input, output, and internal signal, such as std_logic or std_logic_vector.
7. Ensure that each input port, output port, and internal signal is labeled with a unique name.

Figure 3.8 gives an example of such a schematic. Now, from this schematic, we can write a VHDL model using concurrent signal assignment statements. Figure 3.9 shows a template for the VHDL description. This template can be filled in as described in the procedure **Construct_CSA_Model**. Names used in this procedure, such as entity_name, refer to names in the program template of Figure 3.9.

Construct_CSA_Model

1. At this point, I recommend using the IEEE 1164 value system. To do so, include the following two lines at the top of your model declaration:

    library IEEE;
    use IEEE.std_logic_1164.all;

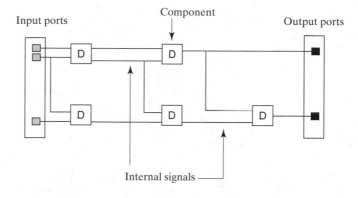

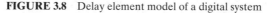

FIGURE 3.8   Delay element model of a digital system

```
library library-name-1, library-name-2;
use library-name-1.package-name.all;
use library-name-2.package-name.all;
entity entity_name is
port( input signals : in type;
        output signals : out type);
end entity entity_name;
architecture arch_name of entity_name is
-- declare internal signals
-- you may have multiple signals of different types
signal internal-signal-1 : type := initialization;
signal internal-signal-2 : type := initialization;
begin
-- specify value of each signal as a function of other signals
internal-signal-1 <= simple, conditional, or selected CSA;
internal-signal-2 <= simple, conditional, or selected CSA;


output-signal-1  <= simple, conditional, or selected CSA;
output-signal-2  <= simple, conditional, or selected CSA;
end architecture arch_name;
```

FIGURE 3.9   A template for writing VHDL models using CSAs

You can declare single-bit signals to be of type std_logic and multibit quantities to
be of type std_logic_vector.

2. Select a name for the entity (entity_name) and write the entity description specifying each input or output signal port, its mode, and associated type. This can be read
   from the annotated schematic.

3. Select a name for the architecture (arch_name) and write the architecture description. Place both the entity and architecture descriptions in the same file. (As we will see in Chapter 8 this is not necessary, in general.)

    3.1 Within the architecture description, name and declare all of the internal signals that connect the components. The declaration states the type of each signal and its initial value. Initialization is not required, but is recommended. These declarations occur prior to the first **begin** statement in the architecture.

    3.2 Each internal signal is driven by exactly one component. If this is not the case, make sure that the type of the signal is a resolved type, such as std_logic or std_logic_vector. For each internal signal, write a concurrent signal assignment statement that expresses the value of this internal signal as a function of input signals for that component. Use the delay value associated with that output signal for that component. This is available from your annotated schematic.

    3.3 Each output port signal is driven by the output of some internal component, that is, each output port is connected to the output of some component. For each output port signal, write a concurrent signal assignment statement that expresses its value as some function of the signals that are inputs to the corresponding component.

    3.4 If you are using any functions or type definitions provided by a third party, make sure that you have declared the appropriate library with the **library** clause and also declared the use of this package via the presence of a **use** clause in your model.

   If there are S signals and output ports in the schematic, there will be S concurrent signal assignment statements in the VHDL model—one for each signal. This approach provides a quick way of constructing VHDL models by attempting to maintain a close correspondence with the hardware being modeled. There are many alternatives for constructing a VHDL model, and the preceding approach represents only one method. With experience, the reader will no doubt discover many other alternatives for constructing efficient models for digital circuits of interest.

Simulation Exercise 3.2: A Single-Bit ALU

Consider a simple one-bit ALU as shown in Figure 3.10 that performs the AND, OR, and ADDITION operations. The result produced at the ALU output depends on the value of signal OPCODE. Write and simulate a model of this ALU, using concurrent signal assignment statements. Test each OPCODE to ensure that the model is accurate by examining the waveforms on the input and output signals. Use a gate delay of 2 ns, a delay of 6 ns through the adder, and a delay of 4 ns through the multiplexor.

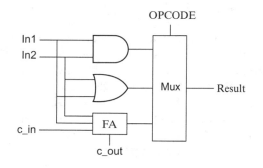

FIGURE 3.10   A single-bit ALU (FA = full adder)

Remember, while the OPCODE field is two bits wide, there are only three valid inputs to the multiplexor.

Step 1. Follow the steps in **Construct_Schematic**. Ensure that all of the signals, including the input and output ports, are defined and labeled and that their mode and types are specified.

Step 2. Follow the steps in **Construct_CSA_Model**. To describe the operation of the full adder, use two simple concurrent signal assignment statements: one each to describe the computation of the sum and carry outputs, respectively. Describe the output of the multiplexor using a conditional signal assignment statement, and use the **when others** or **unaffected** clauses to account for the fact that the multiplexor has only three inputs rather than four. Call this file *alu.vhd*.

Step 3. Compile *alu.vhd*.

Step 4. Generate a sequence of inputs that you can use to verify that the model is functioning correctly.

Step 5. Open a trace window with the signals you would like to trace. Include internal signals, which are signals that are not entity ports in the model.

Step 6. Run the simulation for 50 ns.

Step 7. Check the trace to determine correctness.

Step 8. Print and record the trace.

Step 9. Add new operations to the single-bit ALU, recompile, and resimulate the model. For example, you can add the exclusive-OR, subtraction, and complement operations.

End Simulation Exercise 3.2

3.5   Understanding Delays

We now have a template to help us begin to write basic VHDL models for many digital circuits. Let us examine one important aspect of these models—propagation delays—in greater detail. Accurate representation of the behavior of digital circuits requires accurate modeling of delays through the various components. This section discusses the delay models available in VHDL and how they are specified. These models can be incorporated in a straightforward manner into the basic template for writing VHDL models that was described earlier. They are also used in the models described in Chapter 4 and the structural models described in Chapter 5.

3.5.1   The Inertial Delay Model

In Chapter 2, it was pointed out that digital circuits have a certain amount of inertia. For example, it takes a finite amount of time and a certain amount of energy for the output of a gate to respond to a change on the input. This implies that the change on the input has to persist for a certain period of time to ensure that the output will respond. If it does not persist long enough, the input events will not be propagated to the output. This propagation delay model is referred to as the *inertial delay model* and is the default delay model for VHDL programs.

Figure 3.11 shows an example in which a signal is applied to the input of a 2- input OR gate. If the gate delay is 8 ns, any pulse on the input signal of duration less than 8 ns will not be propagated to the output. This is illustrated by waveform Out 1. However, if the gate delay is 2 ns, we see that each pulse on the input waveform is of a duration greater than 2 ns and is therefore propagated to the output. Any pulse with a width of less than the propagation delay through the gate is said to be rejected. In general, the pulse widths that are actually rejected in a physical circuit depend closely upon the physical design and manufacturing process parameters and can be difficult to determine accurately. The VHDL language uses the propagation delay through the component as the default pulse rejection width.

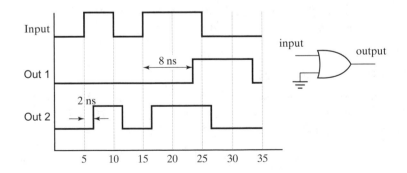

FIGURE 3.11   An example of the inertial delay model. Out 1 is the output waveform for delay = 8 ns and Out 2 is the output waveform for delay = 2 ns

However, if we have a greater understanding of the properties of the components that we are modeling, VHDL supports the following specification of a value for the pulse rejection width:

sum <= **reject** 2 **ns inertial** (x xor y) **after** 5 **ns**;

In Section 3.3.2, a simple form of the waveform element was introduced as the manner in which we could specify the new time–value pair of a signal. The general form of a waveform element allows us to specify a distinct pulse rejection width (distinct from the propagation delay). Note that the expression has been preceded by the keyword **reject**, and a time value has been provided. The keyword **inertial** must also be provided. Thus, in VHDL'93, we can write the general form of the simple concurrent signal assignment statement as follows:

signal <= **reject** *time-expression* **inertial** *value-expression* **after** *time-expression*;

'87 vs. '93

☞ However, VHDL'87 does not support the specification of pulse rejection widths. The delay value is utilized as the pulse rejection width. The preceding statement is a very general form describing the occurrence of an event on a signal. This form specifies the value of the signal, the time at which the signal is to receive this value, and the duration over which the input pulse must persist if the output is to receive this new value.

3.5.2   The Transport Delay Model

Like switching devices, signals propagate through wires at a finite rate and experience delays that are proportional to the distance. However, unlike switching devices, wires have comparatively less inertia. As a result, wires will propagate signals with very small pulse widths, and we can model wires as media that will propagate any changes in signal values independently of the duration of the pulse width. In modern technologies with increasingly small feature sizes the wire delays dominate, and designs seek to minimize wire length. In these circuits, wire delays are nonnegligible and should be modeled to produce accurate simulations of circuit behavior. Such delays are referred to as *transport delays*. While we naturally think of wires as elements with very little inertia, there may be other components that we wish to model for which any event on the input must be propagated to the output. In these cases, we wish to model the behavior of the component with the transport delay model. As with inertial delays, we can specify these delays by prefacing a waveform element with the keyword **transport** as follows:

sum <= **transport** (x xor y) **after** 5 **ns**;

In this case, a pulse of any width on signal x or y can be propagated to the sum signal. We will generally not use the transport delay model for modeling components that have significant inertia. The inertial delay model is the default delay model in VHDL.

Example: Transport Delays

Up to this point, we have treated digital components as delay elements. Output signals acquire values after a specified propagation delay that we now know can be specified to be an inertial delay or a transport delay. If we wish to model delays along wires, we can simply replace the wire with a delay element. The delay value is equal to the delay experienced by the signal transmission along the wire, and the delay type is **transport**. Consider the half-adder circuit again, redrawn to capture wire delays on the output signals, as shown in Figure 3.12. Delay elements model the delay on the sum and carry

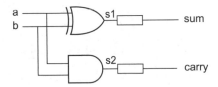

```
library IEEE;
use IEEE.std_logic_1164.all;
entity half_adder is
port(a, b: in std_logic;
     sum, carry: out std_logic);
end entity half_adder;

architecture transport_delay of half_adder is
signal s1, s2: std_logic:= '0';
begin
s1 <= (a xor b) after 2 ns;
s2 <= (a and b) after 2 ns;
sum <= transport s1 after 4 ns;
carry <= transport s2 after 4 ns;
end architecture transport_delay;
```

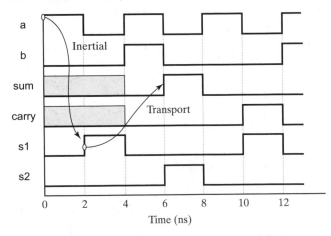

FIGURE 3.12 An example using transport delays

signals. Note in this example that the delay along these wires is longer than the propagation delay through the gate. From the timing diagram, we see that a pulse of width 2 ns on the sum input at time 0 is propagated to signal s1 at time 2 ns. The wire delay is 4 ns. Under the inertial delay model, this pulse would be rejected and would not appear on the sum output. However, we have specified the delay type to be transport. Therefore, the pulse is transmitted to the sum output after a delay of 4 ns. This signal is now delivered to the next circuit, having been delayed by an amount equal to the propagation delay through the signal wires. This approach enables the modeling of wire delays, although, in practice, it is difficult to obtain accurate estimates of the wire delay without proceeding through physical design and the layout of the circuit.

Example End: Transport Delays

The choice between the use of inertial delay or transport delay is determined by the components that are being modeled. For example, if we have a model of a board-level design, we may have VHDL models of the individual chips. The delay experienced by signals between chips can be modeled using the transport delay model. We can modify the procedure **Construct_Schematic** provided in Section 3.4 to include delay elements for all wire delays to be modeled. We simply modify the procedure for translating this annotated schematic so that it uses transport delays in the concurrent signal assignment statements that represent wire delays.

3.5.3   Delta Delays

What happens if we do not specify a delay for the occurrence of an event on a signal? For example, we may write the computation of the outputs for an exclusive-OR gate as follows:

    sum <= (x **xor** y);

We may choose to ignore delays when we do not know what they are or when we are interested only in creating a simulation that is functionally correct and is not concerned with the physical timing behavior. For example, consider the timing of the full-adder model shown in Figure 3.4. There is a correct ordering of events on the signals. Input events on signals In1, In2, and c_in produce events on internal signals s1, s2, and s3, which, in turn, produce events on the output signals sum and c_out. For functional correctness, we must maintain this ordering, even when delays remain unspecified. The VHDL language implementation achieves this by defining an infinitesimally small delay referred to as a *delta delay*. The preceding form of the signal assignment statement implicitly places an "**after** 0 **ns**" time expression following the value expression. When this is the case, the assignment to the signal is effectively assigned a delay value of Δ. Now simulation proceeds exactly as described in the earlier examples using this delay value. As the next example will demonstrate, Δ does not actually have to be assigned a numeric value, but is utilized within the simulator to order events. If events with zero delay are produced at timestep T, the simulator simply organizes and processes events in time order of occurrence: events that occur Δ seconds later are followed by events occurring 2Δ seconds later, followed by events occurring 3Δ seconds later, and so on. Delta delays are simply used to enforce dependencies between events and thereby ensure correct simulation. The next example will help clarify the use of delta delays.

Example: Delta Delays

Consider the combinational logic circuit and the corresponding VHDL code shown in Figure 3.13. The model captures a behavioral description of the circuit *without* specifying any gate delays. Figure 3.14(a) illustrates the timing of the circuit when inputs are applied as shown. At time 10 ns, the signal In2 makes a $1 \rightarrow 0$ transition. This causes a sequence of events in the circuit resulting in the value Z = 0. From the accompanying VHDL code, we see that the gate delays are implicitly 0 ns. Therefore, the timing diagram shows the signal Z acquiring this value at the same instant in time that In2 makes a transition. From the timing diagram, it is also clear that signals s2 and s3 also make

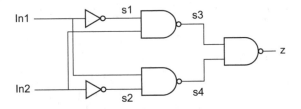

```
library IEEE;
use IEEE.std_logic_1164.all;
entity combinational is
port (In1, In2: in std_logic;
    z : out std_logic);
end entity combinational;

architecture behavior of combinational is
signal s1, s2, s3, s4: std_logic:= '0';
begin
s1 <= not In1;
s2 <= not In2;
s3 <= not (s1 and In2);
s4 <= not (s2 and In1);
z <= not (s3 and s4);
end architecture behavior;
```

FIGURE 3.13   A VHDL model with delta delays

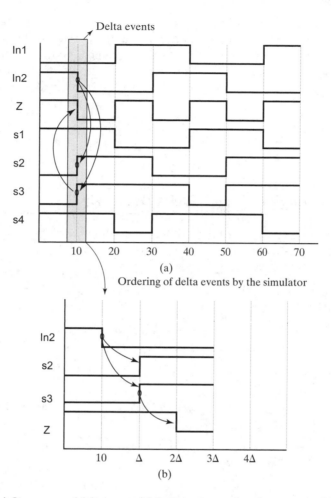

FIGURE 3.14 (a) Occurrence of delta events (b) Ordering imposed on these events within the simulator

transitions at this instant in time. In reality, from the circuit diagram, we know that there is a dependency between In2 and s3, and between s3 and Z. These dependencies are evident from the structure of the circuit. The event on In2 precedes and causes the transitions on s3 and s4, while the transition on s3 causes the transition on Z. The simulation of the circuit honors these dependencies through the logical use of delta delays. The dependencies between In2, s3, and Z are shown in Figure 3.14(b). The transition on In2 causes s3 to make a transition to 1 after Δ secs. The event on s3 causes a $1 \rightarrow 0$ event on Z after 2Δ secs. These events are referred to as delta events and are shown in Figure 3.14(b). These delta events take place within the simulator and do not appear on the external trace produced for the viewer (i.e., Figure 3.14(a)). The actual implementation of delta events is managed within the simulator by keeping track of signal values and when they are updated.

Forcing all events to take some infinitesimally small amount of time preserves the dependencies between events and maintains the correct operation of the circuit. Recall how a circuit is simulated. Time is advanced to that of the first event on the list. The signal is assigned this value, any new outputs are computed, and the process repeats. When time is advanced by Δ, this step is referred to as a delta cycle.

Example End: Delta Delays

Simulation Exercise 3.3: Delta Delays

Repeat the simulation of the full-adder model in Simulation Exercise 3.1, but do not specify any gate delays.

Step 1.   Run the simulation for 40 ns and trace input, internal, and output signals.

Step 2.   Annotate the trace to identify delta events.

Step 3.   Compare the trace generated here with that generated in Simulation Exercise 3.1. What are the differences?

Step 4.   Modify the model to include a 2 ns wire delay for internal signals. Recompile the model.

Step 5.   Simulate the model now and generate another trace. The effect of wire delays is now explicitly captured.

Step 6.   Identify events that occur in this second trace that are different from those in the earlier trace.

Step 7.   Create an input stimulus for one of the inputs with pulses whose duration are both shorter and longer than the gate delay. Set the value of the remaining inputs to logic 0. You can usually achieve this by adjusting the simulator step time and with stimulus commands unique to the simulator that you are using.

Step 8.   Generate a trace and identify pulses that the gate models reject.

End Simulation Exercise 3.3

3.6   Chapter Summary

The reader should be comfortable with the following concepts that have been introduced in this chapter (syntactic reference to common language types and operators can be found in Chapter 9).

- Entity and architecture constructs
- Concurrent signal assignment statements
  - simple concurrent signal assignment
  - conditional concurrent signal assignment
  - selected concurrent signal assignment
- Constructing models using concurrent signal assignment statements
  - modeling events, propagation delays, and concurrency
- Modeling delays
  - inertial delay
  - transport delay
  - delta delay
- Signal drivers and projected waveforms
- Shared signals, resolved types, and resolution functions
- Generating waveforms using waveform elements
- Events and transactions

The VHDL models of systems should now be beginning to take some form. The reader should be capable of constructing functionally correct models for many types of digital circuits utilizing inertial, transport, or delta delay models (i.e., functional models).

Exercises

1. A good exercise for understanding entity descriptions is to write the entity descriptions for components found in data books from component vendors—for example, the TTL data book. These entity descriptions can be compiled without the architecture descriptions, and thus can be checked for syntactic correctness. Of course, we cannot say anything about the semantic correctness of such descriptions, since we have not even written the architecture descriptions yet!

2. Write a VHDL model that generates clocks with frequency of 500 MHz and duty cycles of 0.25, 0.5, and 0.75.

3. Write and simulate a VHDL model of a 2-bit comparator.

4. Sketch the output waveform produced by the following VHDL simple concurrent signal assignment statements:

   s1 <= '0' **after** 5 **ns**, '1' **after** 15 **ns**, '0' **after** 35 **ns**, '1' **after** 50 **ns**;
   s1 <= '0' **after** 20 **ns**, '1' **after** 25 **ns**, '0' **after** 50 **ns**;

5. Construct and test VHDL modules for generating the periodic waveforms with the structure shown in Figure 3.15.

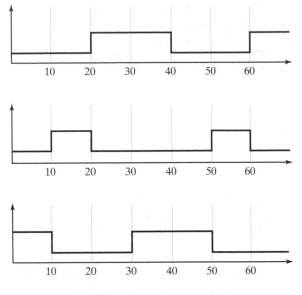

FIGURE 3.15   Sample waveforms

6. Construct a VHDL model that will accept a clock signal as input and produce the complement signal as output with a delay of 10 ns.

7. Construct a VHDL model of a circuit that will produce the following set of non-overlapping clocks as output.

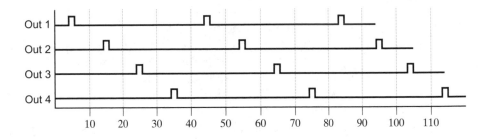

8. Write and simulate the entity–architecture description of a 3-bit decoder using the conditional signal assignment statement. Test the model with all possible combinations of inputs and plot the decoder output waveform.

9. Repeat the preceding exercise by building the decoder from basic gates. Is there any difference in the number of events generated between the simulation of this model and a model that describes behavior by using the conditional signal assignment statement? You should be able to answer this question by examining the traces in both cases over the same time interval. If your simulator permits, examine the event queues during simulation.

10. What is the difference between the hardware behaviors implied by the conditional signal assignment statement and the selected signal assignment statement? For what types of hardware structures would you use one or the other?

11. Write a VHDL model of the following circuit, including wire delays:

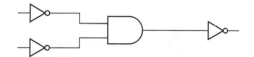

   Use the transport delay model for the wire delays and assume a wire delay of 2 ns between components. Generate a timing diagram. Select your own gate delays. Mark events that would not have occurred under the default inertial delay model.

12. Why are the concepts of delta events and delta delays necessary for the correct discrete event simulation of digital circuits?

13. Write a VHDL model of the gate-level implementation of a four-input priority encoder. Do not specify any gate delays. Identify the points in trace that correspond to delta events.

14. Write and simulate a model of an 8-to-3 priority encoder circuit for which the priority order of inputs is 0, 3, 4, 2, 6, 7, 5, and 1, with 0 being the highest priority input.

15. What delay model would you use if you were interested in functional correctness?

16. Write a VHDL model for a code converter that translates BCD inputs into an excess-3 code.

17. Write a VHDL model of a circuit that accepts a 3-bit binary number as input and generates an output binary number that is the square of the input number.

CHAPTER 4   Modeling Behavior

This chapter expands upon the approach described so far that uses concurrent signal assignment statements to construct VHDL models. In Chapter 3, digital components were modeled as delay elements and their internal behavior was described using concurrent signal assignment statements. Events on input signals caused events on output signals after a propagation delay. This approach works well for modeling digital systems at the gate level. However, it is difficult, and often infeasible with respect to simulation time, to build models of large complex systems at the gate level. We often wish to abstract or hide the details of the hardware implementation while preserving the external event behavior. To be able to write such abstract models requires language constructs more powerful than the concurrent signal assignment statements introduced in the last chapter.

In this chapter, we discuss more powerful constructs for describing the internal behavior of components when they cannot be modeled simply as delay elements. The basis for these descriptions is the *process* construct that enables us to use conventional programming language constructs and idioms. As a result, we can model more complex behaviors than are feasible to model with concurrent signal assignment statements, and we are able to model systems at higher levels of abstraction.

4.1   The Process Construct

The VHDL language and modeling concepts described in Chapter 3 were derived from the operational characteristics of digital circuits, for which the design is represented as a schematic of concurrently operating components. Each component is characterized by the generation of events on output signals in response to events on input signals. These output events may occur after a component-dependent propagation delay. The component behavior is expressed with a CSA statement that explicitly relates input signals, output signals, and propagation delays. Such models are convenient to construct when components correspond to gates or switch-level models of transistors. However, when we wish to construct models of complex components such as CPUs, memory modules, or communication protocols, such a model of behavior can be quite limiting. The event model is still valid—externally we see that events on the input signals will eventually cause events on the output signals of the component. However, the computation of the time at which these output events will occur and the value of the output signals can be quite complex. Moreover, for modeling components such as memories, we need to retain state information within the component description over time. It is not sufficient to be able to compute the values of the output signals as a function of the values of the input signals.

For example, consider the behavior of a simple model of a memory module as shown in Figure 4.1. The memory module is provided with address, data, and read and write control signals. Let us assume that it contains 4096 32-bit words of memory. The values of the MemRead or MemWrite control signals determine whether the data on write_data are to be written at the address on the address port, or whether data are to be read from that address and provided on the output port read_data. Events on the input address, data, or control lines produce events that cause the memory model to be executed. We can also reasonably expect to know the memory access times for read and write operations and therefore know the propagation delays. However, the behavior internal to the memory module is difficult to describe using only the signal assignment statements provided in Chapter 3. How can we represent memory words? How can we address the correct word based on the values of the address lines and control signals? How can memory write operations store values to be subsequently accessed by memory read operations. The answers are easier to come by if we have access to conventional sequential programming language constructs. We can then implement memory as an array and use the address value to index the array. Depending upon the value of the control signals, we can decide whether this array element is to be written or read. We can realize such behavior in VHDL by using *sequential statements* via the *process* construct.

In contrast to concurrent signal assignment statements, a process is a sequentially executed block of code. Figure 4.2 shows a VHDL model of a memory module equivalent to the one in Figure 4.1. This model consists of one process, labeled mem_process. Process labels are delimited by colons. The structure of a process is very similar to that of programs written in a conventional block structured programming language such as

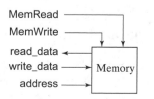

FIGURE 4.1 A model of memory

```
library IEEE;
use IEEE.std_logic_1164.all;
use IEEE.std_logic_arith.all;
                    -- we need this package for 1164 related functions
entity memory is
port(address : in unsigned (31 downto 0); -- used unsigned for memory addresses
    write_data : in std_logic_vector (31 downto 0);
    MemWrite, MemRead : in std_logic;
    read_data : out std_logic_vector (31 downto 0));
end entity memory;

architecture behavioral of memory is
type mem_array is array(0 to 7) of std_logic_vector (31 downto 0);
                    -- define a new type, for memory arrays
begin
mem_process: process (address, write_data) is
variable data_mem : mem_array := ( -- declare a memory array
X"00000000",   --- initialize data memory
X"00000000",   --- X denotes a hexadecimal number
X"00000000",   ---
X"00000000",
X"00000000",
X"00000000",
X"00000000",
X"00000000");
variable addr :integer;

begin

-- the following type conversion function is in std_logic_arith

L1: addr := conv_integer (address (2 downto 0));
L2: if MemWrite = '1' then
L3: data_mem(addr) := write_data; -- perform a read or write operation

elsif MemRead = '1' then
read_data <= data_mem(addr) after 10 ns;
end if;
end process mem_process;
end architecture behavioral;
```

FIGURE 4.2   A behavioral description of a memory module

Pascal. The process begins with a declarative region, followed by the process body, delimited by **begin** and **end** keywords. Variables and constants used within the process are declared within the declarative region. The **begin** keyword denotes the start of the computational part of the process. All of the statements in this process are executed sequentially. Data structures may include arrays and queues, and programs may use standard data types such as integers, characters, and real numbers. Unlike signals, whose changes in values must be scheduled to occur at discrete points in simulated time, variable assignments take effect immediately, just as in conventional programs. Variable assignment is denoted by the ":=" operator. Since all statements are executed sequentially within a process, values assigned to variables are visible to all following statements within that same process. Control flow within a process is strictly sequential, altered by constructs such as **if-then-else** or **loop** statements. In fact, we can regard the process itself as a traditional sequential program. However, one of the distinguishing features of processes is that we can make assignments to signals declared external to the process. For example, consider the memory model in Figure 4.2. At the end of this process, we have signal assignment statements that assign internally (to the process) computed values to signals in the interface after a specified propagation delay. Thus, externally, we are able to maintain the discrete event execution model: Events on the memory inputs produce events on the memory outputs after a delay equal to the memory access time. However, internally we are able to develop complex models of behavior that produce these external events. *With respect to simulation time, a process executes in zero time.* Delays are associated only with the assignment of values to signals.

Recall that a CSA is executed any time an event occurs on a signal in the right-hand side of the signal assignment statement. When is a process executed? In Figure 4.2, adjacent to the **process** keyword is a list of input signals to the component. This is not a parameter list; the list is referred to as the *sensitivity list.* The execution of a process is initiated whenever an event occurs on any of the signals in the sensitivity list of the process. Once started, the process executes to completion in zero (simulation) time and potentially generates a new set of events on output signals. We begin to see the similarity between a process and a CSA. In the models with concurrent signal assignment statements, input signals are inferred by their presence on the right-hand side of the signal assignment statement. In a process, the signals can be listed in the sensitivity list. Remember that the process is sensitive only to signals placed in the sensitivity list. Thus, even if an input signal changes, if this signal is not listed on the sensitivity list, the process will not be executed. For all practical purposes, we can regard a process simply as a "big" concurrent signal assignment statement that executes concurrently with other processes and signal assignment statements. Processes are simply capable of describing more complex events than the CSAs described in Chapter 3. *In fact, CSAs themselves are implemented as processes!* However, they are special and do not require the **process**, **begin**, and **end** syntax of more complex processes.

Several other new language features have made their way into this model. A new type—unsigned—has been introduced in the entity description. The definition of this type is found in the new library package declared in the model: std_logic_arith. Type unsigned is defined as a vector of bits, each of type std_logic, and is often used for objects such as memory addresses. The definition of the type conversion function conv_integer() can also be found in std_logic_arith. This function is necessary because memory is modeled as an array of 32-bit words. This array is indexed by an integer. Therefore, the memory address that is provided as a 32-bit number of type unsigned must be converted to an integer before the array can be accessed. Of course, we do not create an array with 2^{32} entries; rather we form one with eight words of memory. Therefore, the model uses only the lower three bits of the memory address. In this model and a few others in this text, we have used other packages, such as std_logic_arith. Many vendors will support various packages with many useful type conversion, arithmetic, and logic functions. These packages will be placed in various libraries. Check with your installation to determine the location and contents of available packages. This is all we need to know for the moment. We will revisit packages and libraries in greater detail in Chapter 6.

Since statements within a process are executed sequentially, these statements are referred to as *sequential statements,* in contrast to the concurrent signal assignment statements that we saw in Chapter 3. As we see from Figure 4.2, sequential statements can be labeled. For example, note the labels L1, L2, and L3 in the model of Figure 4.2. The use of labels simplifies associated descriptions, as well as improving the clarity of the code. However, individual sequential statements cannot be labeled in VHDL'87. Processes can be thought of as programs that are executed within the simulation to model the behavior of a component. Thus, we have more powerful means to model the behavior of digital systems. Such models are often referred to as behavioral models, although any VHDL model using concurrent or sequential statements is a description of behavior.

'87 vs. '93

☞

Once the concepts of a process and the underlying semantics are understood, we need to know the syntax of the major programming constructs that we can use within a process. Chapter 9 presents identifiers, operators, and useful data types. When we first start developing models, it is easier, although not the most efficient, to write models that are single processes and write them as if we were writing in C or Pascal. On the basis of our experience with other high-level languages, we can begin immediately describing the behavior of components by using processes and start developing nontrivial simulation models. As our experience with the language grows, we will become more selective and effective in how processes are used within larger models. One issue to be kept in mind is that such an approach, although easier, makes the process of synthesis more challenging. In general, I suggest an approach that is predicated on the view that we are describing the behavior of hardware and not describing an algorithm. This will lead to more efficient models while we are climbing the learning curve. It will also naturally lead to design representations in VHDL, making the job of development easier.

4.2   Programming Constructs

This section describes the programming language constructs available for use within a process.

4.2.1   If-Then-Else and If-Then-Elsif Statements

An **if** statement is executed by evaluating a Boolean expression and conditionally executing a block of sequential statements. The structure may optionally include an **else** component. The statement may also include zero or more **elsif** branches. (Note the absence of the second letter 'e' in **elsif**!) In this case, all of the Boolean valued expressions are evaluated sequentially until the first true expression is encountered. An **if** statement is closed by the **end if** clause. The memory model of Figure 4.2 captures a good example of the utility of the use of the **if-then-else** construct.

4.2.2   Case Statement

The behavioral model shown in Figure 4.2 utilizes a single process. Just as we had concurrent signal assignment statements, we may also have concurrently executing processes. Consider another behavioral model of a half adder with two processes, as shown in Figure 4.3. Both processes are sensitive to events on the input signals x and y. Whenever an event occurs on either x or y, both processes are activated and execute concurrently in simulation time. The second process is structured with the use of a

```
library IEEE;
use IEEE.std_logic_1164.all;
entity half_adder is
port (x, y : in std_logic;
      sum, carry : out std_logic);
end entity half_adder;

architecture behavior of half_adder is
begin
sum_proc: process(x,y) is -- this process computes the value of sum
      begin
        if (x = y) then
            sum <= '0' after 5 ns;
        else
            sum <= (x or y) after 5 ns;
        end if;
      end process sum_proc;

carry_proc: process (x,y) is -- this process computes the value of carry
      begin
        case x is
        when '0' =>
        carry <= x after 5 ns;
        when '1' =>
        carry <= y after 5 ns;
        when others =>
        carry <= 'X' after 5 ns;
      end case;
end process carry_proc;
end architecture behavior;
```

FIGURE 4.3   A two-process half-adder model

case statement. The case statement is used whenever it is necessary to select one of several branches of execution based on the value of an expression. The branches of the **case** statement must cover all possible values of the expression being tested. Of course, the choices must be of the same type as that of the value produced by the expression. Each value of the case expression being tested can belong to only one branch of the case statement. The **others** clause can be used to ensure that all possible values for the case expression are covered. Although this example shows a single statement within each branch, in general the branch can be composed of a sequence of sequential statements. The example also shows that port signals are visible within a process. This means that process statements can read port values and schedule values on (i.e., write) output ports.

The preceding example demonstrated that we are not constrained to have a single process within an architecture. In fact, we may even have a mix of processes and concurrent signal assignment statements. Furthermore, just as with selected signal assignment statements in Chapter 3, the select expression can operate on bit fields within a multibit signal. The next example demonstrates these features.

Example: Memory Model

Figure 4.4 shows another model of a memory with four 32-bit words. The model has two interesting features. First, rather than modeling memory as an array of values, it uses four distinct signals. This is clearly not scalable to larger memories and is feasible only for small memories, such as a group of registers, preferably organized as a register file. The second interesting feature of the model is the presence of both a process and a concurrent signal assignment statement in the architecture. The process mem_proc implements the initialization and the memory write operation. This process is sensitive to the clk signal. Whenever there is a change in the value of the clock signal, the process is executed. The function rising_edge (clk) is defined in the package std_logic_1164 and is true if the signal clk has just experienced a rising edge—that is, a change in value from 0 to 1. If so, the reset and MemWrite signals are checked, and the appropriate actions are taken, namely initialization and memory write operations, respectively. This check for the rising edge prevents spurious write operations on the falling edge of clk. The write operation itself is modeled as a **case** statement in which the select expression is the memory address. Note that the input memory address is actually an 8-bit address. However, this particular model implements only four words of memory. The select expression operates on only the least significant two bits of the address. Thus, if a memory address is out of range, the address actually computed will be (address modulo 4). We will find that being able to switch on the values of bit fields within a word is a very useful capability.

Executing concurrently with this process is the memory read operation, which is naturally implemented as a conditional signal assignment statement. We could have included the read operation within mem_proc by extending the **if-then-elsif** construct

```
Library IEEE;
use IEEE.std_logic_1164.all;
use IEEE.std_logic_arith.all;

entity memory is
port (address, write_data : in std_logic_vector (7 downto 0);
MemWrite, MemRead, clk, reset : in std_logic;
read_data :out std_logic_vector (7 downto 0));
end entity memory;

architecture behavioral of memory is
signal dmem0,dmem1,dmem2,dmem3 : std_logic_vector (7 downto 0);
begin
mem_proc: process (clk) is
begin
if (rising_edge(clk)) then -- wait until next clock edge
if reset = '1' then          -- initialize values on reset
dmem0 <= x"00";             -- memory locations are initialized to
dmem1 <= x"11";             -- some random values
dmem2 <= x"22";
dmem3 <= x"33";

elsif MemWrite = '1' then -- if not reset then check for memory write
case address (1 downto 0) is
when "00" => dmem0 <= write_data;
when "01" => dmem1 <= write_data;
when "10" => dmem2 <= write_data;
when "11" => dmem3 <= write_data;
when others => dmem0 <= x"ff";
end case;
end if;
end if;
end process read_proc;

-- memory read is implemented with a conditional signal assignment
read_data <= dmem0 when address (1 downto 0) = "00" and MemRead = '1' else
        dmem1 when address (1 downto 0) = "01" and MemRead = '1' else
        dmem2 when address (1 downto 0) = "10" and MemRead = '1' else
        dmem3 when address (1 downto 0) = "11" and MemRead = '1' else
        x"00";
end architecture behavioral;
```

FIGURE 4.4 An alternative model of memory

with a branch that checked the status of the MemRead control signal. The model would have still operated correctly, although memory read operations would have been synchronous with the clk signal.

Thus, we see that VHDL models can consist of many concurrent constructs within an architecture. Each construct may be a concurrent signal assignment statement or a process.

Example End: Memory Model

4.2.3   Loop Statements

This section introduces the loop statement, as well as a few new VHDL operations that we have not seen before. There are two forms of the loop statement. The first form is the **for** loop. Figure 4.5 shows an example of the use of such a loop construct. This example multiplies two 32-bit numbers by successively shifting the multiplicand and adding to the partial product if the corresponding bit of the multiplier is 1 [13]. Using base 2 arithmetic, the model simply implements what we have traditionally known as long multiplication. The model saves storage by using the lower half of the 64-bit product register to initially store the multiplier. As successive bits of the multiplier are examined, the bits in the lower half of the product register are shifted out, eventually leaving a 64-bit product. Note the use of the **&** operator representing concatenation. This operator can be used to concatenate a k_1-bit word and a k_2-bit word to produce a (k_1+k_2)-bit word. A logical shift right operation is specified by copying the upper 63 (out of 64) bits into the lower 63 bits of the product register and setting the most significant bit to 0 using the concatenation operator. However, in VHDL'93, we can use the built-in shift operators as provided in Chapter 9. VHDL'87 does not support these operators, and one approach towards specifying shift operations is as illustrated in Figure 4.5. Finally, note the inclusion of the package std_logic_unsigned. In the Active VHDL [1] distribution, this package contains the definition for the "+" operator for these operand types.

'87 vs. '93

☞

There are several unique features of this form of a loop statement. Note that the loop variable index is not declared anywhere within the process! The loop index is automatically and implicitly declared by virtue of its use within the loop statement. Moreover, the loop variable index is declared locally for this loop. If a variable or signal with the name index is used elsewhere within the same process or architecture (but not in the same loop), it is treated as a distinct object. Unlike in many conventional programming languages, the loop index cannot be assigned a value or altered in the body of the loop. Therefore, loop indices cannot be provided as parameters via a procedure call or as an input port. We see that the loop index is exactly that—a loop index—and the language prevents us from using it in any other fashion. This does make it convenient to write loops, since we do not have to worry that our choice of variable names for the loop index will conflict with names used elsewhere in the model.

Often, it is necessary to continue the iteration until some condition is satisfied, rather than performing a fixed number of iterations. A second form of the loop statement

```
library IEEE;
use IEEE.std_logic_1164.all;
use IEEE.std_logic_arith.all;
use IEEE.std_logic_unsigned.all; -- needed for arithmetic functions

entity mult32 is
port (multiplicand, multiplier : in std_logic_vector (31 downto 0);
      product : out std_logic_vector (63 downto 0));
end entity mult32;

architecture behavioral of mult32 is
constant module_delay: Time:= 10 ns;
begin
mult_process: process(multiplicand,multiplier) is
variable product_register : std_logic_vector (63 downto 0) := X"0000000000000000";
variable multiplicand_register : std_logic_vector (31 downto 0):= X"00000000";

begin
multiplicand_register := multiplicand;
product_register(63 downto 0) := X"00000000" & multiplier;
--
-- repeated shift-and-add loop
--
for index in 1 to 32 loop
if product_register(0) = '1' then
product_register(63 downto 32) := product_register (63 downto 32) +
multiplicand_register(31 downto 0);
end if;
            -- perform a right shift with zero fill
product_register (63 downto 0) := '0' & product_register (63 downto 1);
end loop;
-- write result to output port
product <= product_register after module_delay;

end process mult_process;
end architecture behavioral;
```

FIGURE 4.5   An example of the use of the loop construct

is the use of the **while** construct. In this form, the **for** statement is simply replaced as in the following example:

```
while j < 32 loop
   ...
   ...
   j := j+1;
   end loop;
```

In general, the statement would appear as "**while** (*condition*) **loop**". Unlike the **for** construct, the loop condition may involve variables that are modified within the loop. Thus, the loop can execute for a data-dependent number of iterations.

4.3   More on Processes

Upon initialization, all processes are executed once. Thereafter, processes are executed in a data-driven manner: activated by events on signals in the sensitivity list of the process or by waiting for the occurrence of specific events using the **wait** statement (described in Section 4.4). Remember, the sensitivity list of a process is not a parameter list! This list simply identifies those signals to which the process is sensitive: When an event occurs on any one of these signals, the process is executed. This operation is analogous to that of CSAs, which are executed whenever an event occurs on a signal on the right-hand side of a CSA. In fact, CSAs are really processes with simpler syntax. Another way of thinking about processes and sensitivity lists is by analogy with combinational logic. When any input signal changes, the network of gates recomputes the values of the output signals. The values of the output signals may not change, but the physical nature of the devices cause them to be recomputed. Processes can be thought of as abstractions of such circuits and thus are executed when any of the signals in the sensitivity list change.

The fact that VHDL is a hardware description language often produces effects that may be unexpected. For example, consider successive signal assignment statements within a process. The behavior you might expect is not what is defined by the language. The next example illustrates this point.

Example: Signal Assignments within Processes

The example presented in Figure 4.6 shows two processes, proc1 and proc2, each of which executes the same sequence of operations in computing the value of the output signals res1 and res2, respectively. However, the computation of res1 uses intermediate variables var_s1 and var_s2, while the computation of the value of res2 utilizes intermediate signals sig_s1 and sig_s2. This example specifies no delays for the signal assignment statements. Therefore the delta delay model is used to order events to preserve correctness. Consider the execution of proc1. The statement L1 computes the value of var_s1, which statement L2 uses to compute the current value of var_s2. Statement L3 then uses the new values of var_s1 and var_s2 computed at the current time to compute the value of signal res1. Now consider the execution of proc2. Statements L1 and L2 compute new values of sig_s1 and sig_s2. These signals do not acquire those new values until the next simulation cycle! This means that the process must run to completion using the existing values of sig_s1 and sig_s2. Thus, the execution of statement L1 will produce a future value of sig_s1 scheduled to take place 0 ns and one delta delay into the future.

```
library IEEE;
use IEEE.std_logic_1164.all;

entity sig_var is
port (x, y, z: in std_logic;
res1, res2 : out std_logic);
end entity sig_var;

architecture behavior of sig_var is
signal sig_s1, sig_s2 : std_logic;
begin
proc1: process (x, y, z) is -- Process 1
variable var_s1, var_s2: std_logic;
begin
L1: var_s1 := x and y;
L2: var_s2 := var_s1 xor z;
L3: res1 <= var_s1 nand var_s2;
end process;
proc2: process (x, y, z) -- Process 2
begin
L1: sig_s1 <= x and y;
L2: sig_s2 <= sig_s1 xor z;
L3: res2 <= sig_s1 nand sig_s2;
end process;
end architecture behavior;
```

FIGURE 4.6   An example of using signals in a process

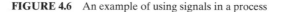

(See Section 3.5.3.) The execution of statement L2 will use the current value of sig_s1 and will compute a new value of sig_s2 that will be assigned to sig_s2 at 0 ns and one delta delay in the future. The computation of res2 will use the current, and not the newly computed, values of sig_s1 and sig_s2. Figure 4.7 illustrates the effect of this behavior. Note that, other than the use of signals vs. variables, the values of res1 and res2 are computed by using the same set of equations. Now observe the waveforms: They are distinct! A good rule of thumb is to keep in mind that you are modeling hardware. Objects that represent signals in the modeled system should be represented as signals. Objects that are used simply to compute values of signals can naturally use variables.

When you are writing and debugging VHDL programs, we must remember to think of these signals as hardware entities and that processes are describing a sequence of computations to determine values that are to be assigned to signals. These assignments do not take place until the process has finished executing. On the other hand, variables act just as we expect from conventional programming languages.

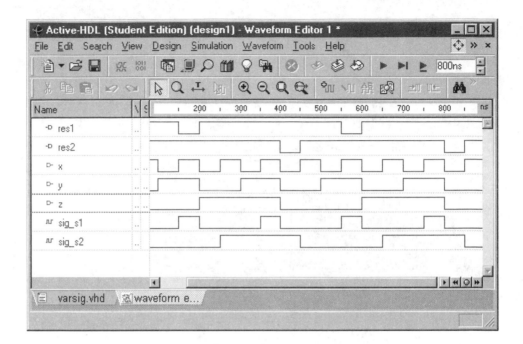

FIGURE 4.7   Illustration of the difference between signals and variables

---

Example End: Signal Assignments within Processes

Another aspect of processes concerns the visibility of signals. All of the ports of the entity and the signals declared within an architecture are visible within a process, which means that the process can read them or assign values to them. Thus, during the course of its execution, a process may read or write any of the signals declared in the architecture or any of the ports on the entity. This is how processes can communicate among themselves. For example, process A may write a signal that is in the sensitivity list of process B. This will cause process B to execute. Process B may in turn similarly write a signal in the sensitivity list of process A. The use of communicating processes is elaborated on in the next example.

Example: Communicating Processes

This example illustrates a model of a full adder constructed from two half adders and a two-input OR gate. Figure 4.8 shows the behaviors of the three components, which are described with processes that communicate through signals. When there is an event on

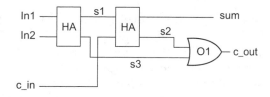

```
library IEEE;
use IEEE.std_logic_1164.all;

entity full_adder is
port (In1, c_in, In2 : in std_logic;
     sum, c_out : out std_logic);
end entity full_adder;

architecture behavior of full_adder is
signal s1, s2, s3: std_logic;
constant delay: Time:= 5 ns;
begin
HA1: process (In1, In2) is --process describing the first half adder
begin
s1 <= (In1 xor In2) after delay;
s3 <= (In1 and In2) after delay;
end process HA1;

HA2: process (s1, c_in) is --process describing the second half adder
begin
sum <= (s1 xor c_in) after delay;
s2 <= (s1 and c_in) after delay;
end process HA2;

OR1: process (s2, s3) is --process describing the two-input OR gate
begin
c_out <= (s2 or s3) after delay;
end process OR1;
end architecture behavioral;
```

FIGURE 4.8   A communicating process model of a full adder

either input signal, process HA1 executes, creating events on internal signals s1 and s2. These signals are in the sensitivity lists of processes HA2 and O1; therefore, these processes will execute and schedule events on their outputs as necessary. Note that this style of modeling still follows the structural description of the hardware with one process for each hardware component of Figure 4.8. Contrast this model with the model described in Figure 3.3.

Example End: Communicating Processes

Simulation Exercise 4.1: Combinational Shift Logic

This exercise is concerned with the construction of a combinational logic shifter shown in Figure 4.9. The inputs to the shift logic include a 3-bit operand specifying the shift amount, two single-bit signals identifying the direction of the shift operation—left or right—and an 8-bit operand. The output of the shift logic is the shifted 8-bit operand. These shift operations are logical shifts and therefore provide zero fill. For example, a left shift of the number 01101111 by 3 bit positions will produce the output 01111000.

Step 1. Create a text file with the entity description and the architecture description of the shift logic. Assume that the delay through the shift logic is fixed at 40 ns, independently of the number of digits that are shifted. While you can implement this behavior in many ways, for this assignment use a single process and the sequential VHDL statements to implement the behavior of the shift logic. You might find it useful to use the concatenation operator **&** and addressing within arrays to perform the shift operations. For example, we can write the following assignment:

dataout <= datain(4 **downto** 0) & "000";

This assignment statement will perform a left shift by three digits with zero fill. Both input and output operands are 8-bit numbers. For VHDL'93, you may use the VHDL built-in shift operators. Use the case statement to structure your process.

Step 2. Use the types std_logic and std_logic_vector for the input and output signals. Declare and reference the library IEEE and the package std_logic_1164.

Step 3. Create a sequence of test vectors. Each of the test vectors will specify the values of (1) the shiftright and shiftleft single-bit control signals, (2) an 8-bit input operand, and (3) a 3-bit number that specifies the number of digits the input operand is to be shifted. Your test cases should be sufficient to ensure that the model is operating correctly.

Step 4. Load the simulation model into the simulator. Set the simulator step time to be equal to the value of the propagation delay through the shift logic.

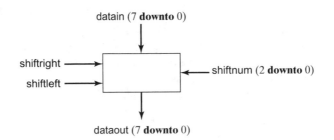

FIGURE 4.9   Interface description for a combinational logic shifter

Step 5. Using the facilities available within the simulator, generate the input stimulus and open a trace window to view both the input stimulus and the output operand value.

Step 6. Exercise the simulator by running the simulation long enough to cover your test cases. Verify correct operation from the trace.

Step 7. Once you have the simulation functioning correctly, modify your model to implement circular shift operations. These operations are such that the digits shifted out of one end of the operand are the inputs to the other end. For example, a circular left shift of the pattern 10010111 by three digits will be 10111100. The circular shift operations can be implemented using the concatenation operator. In VHDL'93, the circular shift operations can be implemented with the VHDL predefined operators.

'87 vs. '93
☞

End Simulation Exercise 4.1

4.4 The Wait Statement

The execution behavior of the models presented in Chapter 3 and the behavioral models described so far in this chapter have been data driven, wherein events on the input signals initiated the execution of concurrent signal assignment statements or processes. Signal assignment statements suspend execution until the next event on a signal on the RHS of the assignment statement. Each process will then suspend execution until the next event on a signal defined in its sensitivity list. This behavior fits in well with the behavior of combinational circuits, in which a change on the input signals may cause a change in the value of the output signals. Therefore, the outputs should be recomputed whenever there is a change in the value of the input signal.

However, what about modeling circuits for which the outputs are computed only at specific points in time, independently of events on the inputs? How do we model circuits that respond only to certain events on the input signals? For example, in synchronous sequential circuits, the clock signal determines when the outputs may change or when inputs are read. Such behavior requires us to be able to specify in a more general manner the conditions under which the circuit outputs must be recomputed. In VHDL terms, we need a more general way of specifying when a process is executed or suspended pending the occurrence of an event or events. This capability is provided by the **wait** statement.

The **wait** statement explicitly specifies the conditions under which a process may resume execution after being suspended. The forms of the **wait** statement include the following:

> **wait for** *time expression*;
> **wait on** *signal*;
> **wait until** *<condition>*;
> **wait**;

The first form of the wait statement causes suspension of the process for a period of time given by the evaluation of *time expression*. This is an expression that should evaluate to a value that is of type **time**. The simplest form of this statement is as follows:

wait for 20 **ns**;

The second form causes a process to suspend execution until an event occurs on one or more signals in a group of signals. For example, we might have the following statement:

wait on clk, reset, status;

In this case, an event on any of the signals causes the process to resume execution with the first statement following the **wait** statement. The third form can specify a *<condition>* that evaluates to a Boolean value, TRUE or FALSE.

Using these wait statements, processes model components that are not necessarily data driven, but are driven only by certain types of events, such as the rising edge of a clock signal. Many such conditions cannot be described with sensitivity lists alone. More importantly, we would often like to construct models in which we need to suspend a process at multiple points within the process and not just at the beginning. Such models are made possible through the use of the **wait** statement. *Because of this behavior of **wait** statements, a process can have a sensitivity list or use **wait** statements, but not both!* The next few examples will help further motivate the use of the **wait** statement.

Example: Positive Edge-Triggered D Flip-Flop

The model of a positive edge-triggered D flip-flop is a good example of the use of the wait statement. The behavior of this component is such that the D input is sampled on the rising edge of the clock and transferred to the output. Therefore, the model description must be able to specify computations of output values only at specific points in time—in this case, the rising edge of the clock signal. This is done with the use of the **wait** statement, as shown in Figure 4.10. This brings us to another very interesting feature of the language. Note the statement clk'**event** in the model shown in the figure. This statement is true if an event (i.e., signal transition) has occurred on the clk signal. The conjunction (clk'**event and** clk = '1') is often used to detect a rising edge on the clk signal. The signal clock is said to have an *attribute* named **event** associated with it. The predicate clk'**event** is true whenever an event has occurred on the signal clk in the most recent simulation cycle. Recall that an event is a change in the signal value. In contrast, a *transaction* occurs on a signal when a new assignment has been made to the signal, but the value may not have changed. As this example illustrates, such an attribute is very useful. The next section lists some useful attributes of VHDL objects.

The std_logic_1164 package also provides two functions that we could have used in lieu of the **event** attribute: rising_edge (clk) and falling_edge (clk). These functions take a signal of type std_logic as an argument and return a Boolean value denoting

```
library IEEE;
use IEEE.std_logic_1164.all;
entity dff is
port (D, Clk : in std_logic;
      Q, Qbar : out std_logic);
end entity dff;

architecture behavioral of dff is
begin
output: process is
begin
wait until (Clk'event and Clk = '1'); -- use the function rising_edge(Clk)
                                       -- for true rising edge detection
   Q <= D after 5 ns;
   Qbar <= not D after 5 ns;

end process output;
end architecture behavioral;
```

FIGURE 4.10   Behavioral model of a positive edge-triggered D flip-flop

whether a rising edge (falling edge) occurred on the signal. The predicate clk'**event** simply denotes a change in value. Note that a single-bit signal of type std_logic can have up to nine values. Thus, if we are really looking for a rising edge from signal value 0 to 1, or a falling edge from signal value 1 to 0, it would be better to replace the test "**if** (Clk'**event and** Clk = '1')" with "**if** rising_edge(Clk)," since the former will be true even if Clk transitions from, say, X to 1.

Continuing with the description of the operation of the D flip-flop, we see that the input is sampled on the rising clock edge and the output values are scheduled after a period equal to the propagation delay through the flip-flop. The process is not executed whenever there is a change in the value of the input signal D, but rather only when there is a rising edge on the signal Clk. Thus, outputs are computed only on rising clock edges, just as in the operation of a physical implementation.

Example End: Positive Edge-Triggered D Flip-Flop

The preceding example did not specify the initial values of the flip-flop. When a physical system is powered up, the individual flip-flops may be initialized to some known state, but not necessarily all in the same state. In general, it is better to have some control over initial states of the flip-flops. This is usually achieved by providing such inputs as Clear or Set and Preset or Reset. Asserting the Set input forces Q = 1 and asserting the Reset input forces Q = 0. These signals override the effect of the clock signal and are active at any time—hence the characterization as asynchronous inputs, as opposed to the synchronous nature of the clock signal. The next example illustrates how we can extend the previous model to include asynchronous inputs.

Example: D Flip-Flop with Asynchronous Inputs

Figure 4.11 shows a model of a D flip-flop with asynchronous reset (R) and set (S) inputs and the corresponding VHDL model. The R input overrides the S input. Both signals are active low. Therefore, to set the output Q = 0, a zero pulse is applied to the reset input while the set input is held to 1, and vice versa. During synchronous operation, both S and R must be held to 1.

```vhdl
library IEEE;
use IEEE.std_logic_1164.all;
entity asynch_dff is
port (R, S, D, Clk : in std_logic;
     Q, Qbar : out std_logic);
end entity asynch_dff;

architecture behavioral of asynch_dff is
begin
output: process (R, S, Clk) is
begin
if (R = '0') then
    Q <= '0' after 5 ns;
    Qbar <= '1' after 5 ns;
elsif S = '0' then
    Q <= '1' after 5 ns;
    Qbar <= '0' after 5 ns;
   elsif (rising_edge(Clk)) then
    Q <= D after 5 ns;
    Qbar <= (not D) after 5 ns;
end if;
end process output;
end architecture behavioral;
```

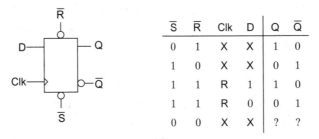

$\overline{S}$	$\overline{R}$	Clk	D	Q	$\overline{Q}$
0	1	X	X	1	0
1	0	X	X	0	1
1	1	R	1	1	0
1	1	R	0	0	1
0	0	X	X	?	?

FIGURE 4.11   D flip-flop with asynchronous set and reset inputs

Example End: D Flip-Flop with Asynchronous Inputs

Now that we have seen how to create a model for a basic unit of storage with asynchronous inputs, it is relatively straightforward to create models for registers and counters in a similar manner. The next example does just that.

Example: Registers and Counters

We can construct a model of a typical 4-bit register composed of edge-triggered D flip-flops, with asynchronous clear and enable signals. Such a model is shown in Figure 4.12. With a few modifications, this example can be converted into a model of a counter. On each clock edge, rather than sampling the inputs, we can simply increment the value stored in the register. The initialization step can be also changed to load a preset value into the counter rather than initializing the counter to 0.

```vhdl
library IEEE;
use IEEE.std_logic_1164.all;
entity reg4 is
port (D : in std_logic_vector (3 downto 0);
      Cl, enable, Clk: in std_logic;
      Q : out std_logic_vector (3 downto 0));
end entity reg4;

architecture behavioral of reg4 is
begin
reg_process: process (Cl, Clk) is
begin
if (Cl = '1') then
    Q <= "0000" after 5 ns;
  elsif (rising_edge(Clk)) then
  if enable = '1' then
  Q<= D after 5 ns;
  end if;
end if;
end process reg_process;
end architecture behavioral;
```

FIGURE 4.12 A 4-bit register with asynchronous inputs and enable signals

Example End: Registers and Counters

Example: Asynchronous Communication

Another example of the utility of wait statements is the modeling of asynchronous communication between two devices. Figure 4.13 shows a simple four-phase protocol for synchronizing the transfer of data between a producer and consumer. Let us assume that the producer (e.g., an input device such as a microphone) is providing data for a consumer device (e.g., the processor) and that the transfer of each word must be synchronized. When the producer has data to be transferred, the signal RQ is asserted. The consumer waits for a rising edge of the RQ signal before reading the data. The consumer then signals successful reception of the data by asserting the ACK signal. This causes the producer to de-assert RQ, which in turn results in the consumer de-asserting ACK. At this point, the transaction is completed, and the consumer and producer can each assert that the other has successfully completed its end of the transaction. The producer and consumer can be modeled as processes that communicate via signals. Such a model is shown in Figure 4.14. Note that signals RQ, ACK, and transmit_data are declared in the architecture and are visible within both processes. Although this example is incomplete in that the processes do not perform any interesting computations with the data, it does illustrate the use of wait statements to control asynchronous communication between processes. Moreover, it illustrates the ability to suspend the execution of a process at multiple points within the model. Since these processes execute concurrently in simulated time, they must be capable of suspending and resuming execution at multiple points within the VHDL code. Such a model is not possible using only sensitivity lists as the mechanism for initiating process execution (although other solutions to the producer–consumer problem are possible).

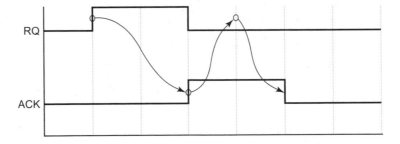

FIGURE 4.13   Four-phase handshake

```vhdl
library IEEE;
use IEEE.std_logic_1164.all;

entity handshake is
port (input_data : in std_logic_vector(31 downto 0));
end entity handshake;

architecture behavioral of handshake is
signal transmit_data: std_logic_vector (31 downto 0);
signal RQ, ACK : std_logic;
begin
producer: process is
begin
wait until input_data'event; -- wait until input data are available
transmit_data <= input_data;    -- provide data as producer
RQ <= '1';
wait until ACK = '1';
RQ <= '0';
wait until ACK = '0';
end process producer;

consumer: process is
variable receive_data : std_logic_vector (31 downto 0);
begin
wait until RQ = '1';
receive_data := transmit_data; -- read data as consumer
ACK <= '1';
wait until RQ = '0';
ACK <= '0';
end process consumer;
end architecture behavioral;
```

FIGURE 4.14   A VHDL model of the behavior shown in Figure 4.13

Example End: Asynchronous Communication

4.5   Attributes

The example of the model of the D flip-flop introduced the idea of an attribute of a
signal. Attributes can be used to formulate a query about a signal to determine various
types of information about the signal. For example, we have seen that the predicate

clk'**event** returns true if there has been a change in the value of the signal during the last simulation cycle. This particular attribute, **event**, causes a function to be called that will return TRUE or FALSE. What other types of information can be obtained about a signal? What about obtaining information about other VHDL names such as entities or arrays? This notion of attributes appeals to our intuition, as just about any object possesses attributes. For example, a car has attributes of color, size, power, and cost. The notion of attributes, as provided in VHDL, is quite general. The language pre-defines some attributes, while the programmer can create user-defined attributes. In this section, we will limit our presentation to commonly used predefined attributes.

The language predefines five basic types of attributes:

- *Function attributes*: invoke a function which returns a value.
- *Value attributes*: return a constant value.
- *Signal attributes*: return a signal.
- *Type attributes*: return a type.
- *Range attributes*: return a range.

4.5.1   Function Attributes

Attributes such as **event**, which are applied to signals, are referred to as *function attributes*—a function is called that returns a value. Table 4.1 shows some other useful function attributes for signals. For example, in addition to determining whether an event has occurred on the signal, we might be interested in knowing the amount of time that has elapsed since the last event occurred on the signal. This attribute is denoted with the use of the following syntax:

clk'**last_event**

In effect, when the simulator executes the preceding statement, a function call occurs that checks this property. The function returns the time since the last event occurred on signal clk.

TABLE 4.1   Some useful function signal attributes

Function attribute	Function
signal_name'**event**	Function returning a Boolean value signifying a change in value on this signal
signal_name'**active**	Function returning a Boolean value signifying an assignment made to this signal. This assignment may not be a new value.
signal_name'**last_event**	Function returning the time since the last event on this signal
signal_name'**last_active**	Function returning the time since the signal was last active
signal_name'**last_value**	Function returning the previous value of this signal
my_array'**length**	Function returning the length of the array my_array

4.5.2   Value Attributes

A second class of predefined attributes is *value attributes*. As the name suggests, these attributes return constant values. For example, the memory model shown in Figure 4.2 contains the definition of a new type as follows:

type mem_array **is array**(0 **to** 7) **of** std_logic_vector (31 **downto** 0);

From Table 4.2, we have mem_array'**left** = 0, and mem_array'**ascending** = true and mem_array'**length** = 8. Another good example involves the use of enumerated types. For instance, when writing models of state machines described later in this chapter, it is useful to have the following data type defined:

type statetype **is** (state0, state1, state2, state3);

This statement defines a new datatype called statetype. Any variable or signal declared to be of that type can take on one of four values: state0, state1, state2, or state3. In the current case, we have statetype'**left** = state0 and statetype'**right** = state3. This is useful in behavioral models when we wish to initialize signals to values on the basis of their types. We may not always know the range and values of the various data types, nor would we really care to remember. The use of attributes makes it easy to initialize objects to values without having to be concerned with the implementation. For example, on a reset operation, we may simply initialize a state machine to the leftmost state of the enumerated list of possible states—that is, statetype'**left**. Some commonly used value attributes are shown in Table 4.2.

TABLE 4.2   Some useful value attributes

Value attribute	Value
scalar_name'**left**	returns the leftmost value of scalar_name in its defined range of values
scalar_name'**right**	returns the rightmost value of scalar_name in its defined range of values
scalar_name'**high**	returns the highest value of scalar_name in its range of values
scalar_name'**low**	returns the lowest value of scalar_name in its range of values
scalar_name'**ascending**	returns true if scalar_name has an ascending range of values (VHDL'93 only)
array_name'**length**	returns the number of elements in the array array_name
array_name'**left**	the left bound of array_name
array_name'**right**	the right bound of array_name

'87 vs. '93

4.5.3   Signal Attributes

Signal attributes create new signals from the signals that you explicitly declare in VHDL models. These new signals are referred to as *implicit* signals. For example, the attribute signal_name'**delayed(T)** creates a new signal of the same type as signal_name, but which is delayed by T. If no delay is specified, then a delta delay is inserted between the two signals.

Several such types of implicit signals can be created in VHDL as shown in Table 4.3. The '**transaction** attribute creates a signal that toggles when the original signal changes value. The '**stable** and '**quiet** attributes are signals that provide for easy tests of the behavior of signals over periods of time.

Keep in mind that these implicit signals are actually generated and can be used elsewhere in the model. For example, consider the simple code block shown in Figure 4.15, where signal attributes create two implicit signals. The associated entity

TABLE 4.3   Some useful signal attributes

Signal attribute	Implicit Signal
signal_name'**delayed(T)**	Signal delayed by T units of time
signal_name'**transaction**	Signal whose value toggles when signal_name is active
signal_name'**quiet(T)**	True when signal_name has been quiet for T units of time
signal_name'**stable(T)**	True when event has not occurred on signal_name for T units of time

```
architecture behavioral of attributes is
begin
    outdelayed <= data'delayed(5 ns);
    outtransaction <= data'transaction;
end attributes;
```

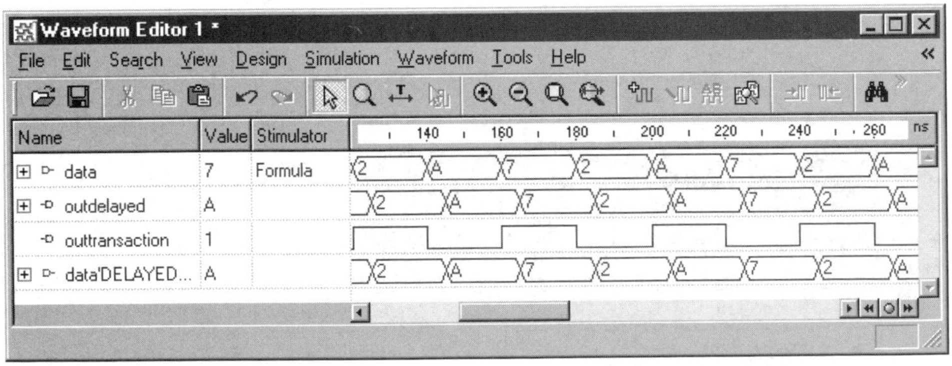

FIGURE 4.15   An example of the use of signal attributes and implicit signals

description (not shown) has a signal data as a 4-bit input signal, and the architecture produces two output signals derived from data, using signal attributes. The first is a 4-bit signal delayed from data by 5 ns. The second signal corresponds to data'**transaction**. This implicit signal toggles whenever the signal data changes value.

When would one use an implicit signal? If we are interested in waiting for a change in value on a signal, we can simply wait for events on the '**transaction** implicit signal. Furthermore, the presence of a delayed signal can be used to check for relationships between the current value of the signal and an older value of the same signal. For example, we might wish to check whether the change in value has been greater than a certain amount:

> **wait on** ReceiveData'**transaction**
>> **if** ReceiveData'**delayed (5 ns)** = ReceiveData **then**
>>> ......

4.5.4   Range Attribute

Finally, a very useful attribute of arrays is the **range** attribute. This is useful in writing loops. For example, consider a loop that scans all of the elements in an array value_array(). The attribute given by value_array'**range** returns the index range. This makes it very easy to write loops, particularly when we may not know the size of the array, as is sometimes the case when writing functions or procedures for which the array size is determined when the function is called. When we do not know the array size, we can write the loop as follows:

> **for** i **in** value_array'**range loop**
> ...
> my_var := value_array(i);
> ...
> **end loop**;

Some specific examples of the use of the range attribute can be found in the discussion of functions and procedures in Chapter 6.

4.5.5   Type Attributes

As the name suggests, this class of attributes enables queries about the type of names used in VHDL programs. We will relegate this to the list of advanced topics that are not dealt with here.

4.6   Generating Clocks and Periodic Waveforms

Since wait statements provide the programmer with explicit control over the reactivation of processes, they can be used to generate periodic waveforms, as shown in the next example.

Example: Generating Periodic Waveforms

We know that, in a signal assignment statement, we can specify several future events. For example, we might have the following signal assignment statement:

signal <= '0', '1' **after** 10 **ns**, '0' **after** 20 **ns**, '1' **after** 40 **ns**;

The execution of this statement will create the waveform shown in Figure 4.16. Now, if we place the statement within a process and use a wait statement, we can cause the process to be executed repeatedly, producing a periodic waveform. Recall that, upon initialization of the VHDL model, all processes are executed. Therefore, every process is executed at least once. During initialization, the first set of events shown in the waveform will be produced. The execution of the **wait for** statement causes the process to be reactivated after 50 ns. This will cause the process to be executed again, generating events in the interval 50–100 ns. The process again suspends execution for 50 ns, and the cycle is repeated. By altering the durations in the statement in the process, one can envision the generation of many different types of periodic waveforms. For example, if we wished to generate two-phase nonoverlapping clocks, we could utilize the same approach, as shown in the next example.

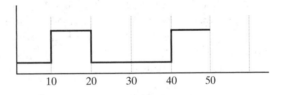

```
library IEEE;
use IEEE.std_logic_1164.all;

entity periodic is
port (Z : out std_logic);
end entity periodic;

architecture behavioral of periodic is
begin
process is
begin
Z <= '0', '1' after 10 ns, '0' after 20 ns, '1' after 40 ns;
wait for 50 ns;
end process;
end architecture behavioral;
```

FIGURE 4.16   An example of the generation of periodic waveforms

Example End: Generating Periodic Waveforms

Example: Generating a Two-Phase Clock

Figure 4.17 gives an example of a model for the generation of nonoverlapping clocks and reset pulses. Such signals are very useful and are found in the majority of circuits we will come across. The reset process is a single concurrent signal assignment statement; therefore, we can dispense with the **begin** and **end** statements. Recall that CSAs are processes and we can assign them labels. Every process is executed just once, at initialization. During this initialization, reset is executed, generating a pulse of width 10 ns. Since there are no input signals, the reset statement is never executed again! The clock process, on the other hand, generates multiple clock edges in a 20-ns interval with each statement. Note the width of the pulses in the second clock signal: It is adjusted so as to prevent the pulses from overlapping. The **wait for** statement causes the process to be executed again 20 ns later, when each statement generates clock edges in the next 20-ns interval. The process repeats indefinitely, generating the waveforms shown in Figure 4.17. Note how concurrent signal assignment statements

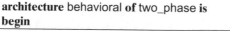

```
library IEEE;
use IEEE.std_logic_1164.all;
entity two_phase is
port (phi1, phi2, reset : out std_logic);
end entity two_phase;

architecture behavioral of two_phase is
begin
reset_process: reset <= '1', '0' after 10 ns;          Generate a
                                                        reset pulse

clock_process: process is
begin
phi1 <= '1', '0' after 10 ns;                           Clock process
phi2 <= '0', '1' after 12 ns, '0' after 18 ns;
wait for 20 ns;
end process clock_process;
end entity behavioral;
```

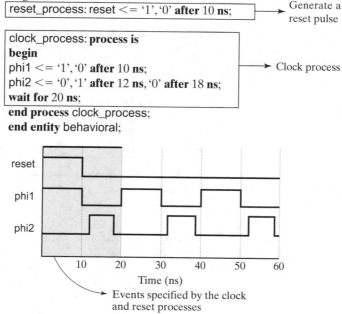

FIGURE 4.17   Generation of two-phase nonoverlapping clocks

are mixed in with the process construct. This type of modeling, using both concurrent and sequential statements, is quite common, which is not surprising, since CSAs are essentially processes. If we take the viewpoint that all statements in VHDL are concurrent, then processes using sequential statements can be viewed as one complex signal assignment statement. This is a useful template to have in mind when constructing behavioral models of hardware.

Example End: Generating a Two-Phase Clock

4.7   Using Signals in a Process

We can think of processes as conventional programs that can be used in VHDL simulation models to provide us with powerful techniques for the computation of events. However, the sequential nature of processes, in conjunction with the use of signals within a process, can produce behavior that may be unexpected. For example, consider the circuit and corresponding model and timing behavior illustrated in Figure 3.13 and Figure 3.14. Now, let us enclose the concurrent signal assignment statements shown in the model of Figure 3.13 in a process sensitive to the input signals. This would represent a different model of operation for the following reason: The circuit being described is a combinational circuit; signal values are determined in a data-driven manner; signal assignment statements are executed only when input signal values change. Thus, the value of signal s3 is computed only when there is a transaction on signals s1 or In2.

Now, consider the implementation using processes for which the timing is shown in Figure 4.18. When there is a change in value on either In1 or In2, the process is executed. By definition, a process executes to completion. All statements within a process are executed! Consider the value of signal s3 at initialization time when each process is

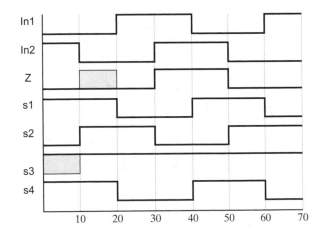

FIGURE 4.18   The effect of processes on signal assignment statements

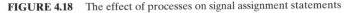

executed. The value of this signal is undefined. Signal s4 has the value 1, since In1 is 0. Therefore, the value of s4 is 1, regardless of the value of s2. However, the value of s3 is undefined and it appears as such on the trace. Now, the process suspends execution and waits for an event on In1 or In2. It will not be executed again until 10 ns later, when a $1 \rightarrow 0$ transition occurs on In2. From the trace in Figure 4.18, we can see that forcing the process statements to be executed in order and being sensitive only to the inputs In1 and In2 produces a trace very distinct from that in Figure 3.14. Although both models were intended to represent the same circuit, we have realized different behaviors. One could make a good case that the use of a process places artificial constraints on the evaluation of signal values and that the former model is indeed a more accurate reflection of the physical behavior of the circuit. In reality, if we made the process sensitive to all of the signals (i.e., including s1, s2, s3, and s4), we would find that the process model would behave exactly as the earlier model producing an identical trace. However, this is not a very intuitive description of the model. Suffice to say that when using signals within a process, one must be careful that the behavior that results is indeed what the modeler had in mind.

Simulation Exercise 4.2: Use of Wait Statements

This simulation exercise provides an introduction to the use of wait statements to suspend and resume processing under programmer control. We can use such constructs to respond to asynchronous external events. Figure 4.19 shows an example of a simple interface that reads data from a device and buffers the data for an output device. Let us model this interface with two processes. The first process communicates with the second via a handshaking protocol such as the one demonstrated in Figure 4.13. The second process can then drive an external device such as a display.

Step 1. Using a text editor, construct a VHDL model for communication between an input process and an output process, using the handshaking protocol captured in Figure 4.13. Assume that the input process can read only a single word at a time. The input process receives a single 32-bit word composed of four bytes. This word is to be transferred to an output device whose storage layout requires reversing the byte order within the word. This reversal is performed by the input process before it is transferred to the output process, which in turn writes the value to an output port.

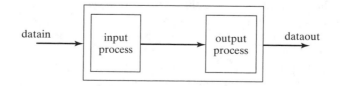

FIGURE 4.19 An example of asynchronous communication

Step 2. Use the types std_logic and std_logic_vector for input and output signals. Declare and reference the library IEEE and package std_logic_1164.

Step 3. Create a sequence of 32-bit words as test inputs.

Step 4. Compile the model and load the compiled model into the simulator.

Step 5. Assign a delay of 10 ns for each handshake transition.

Step 6. Open a trace window and select the signals to be traced. Since we are dealing with 32-bit quantities, set the trace to display these signal values in hexadecimal notation. This will make it considerably easier to read the values in the trace. Such commands are simulator specific.

Step 7. Simulate for several hundred nanoseconds.

Step 8. Trace the four-phase handshake sequence of Figure 4.13.

Step 9. From the trace, determine how long it takes for the input process to transfer a data item to the output process.

Step 10. Does the rate at which the input items are provided matter? What happens as we increase the frequency with which data items are presented to the input process? What is the maximum input data rate?

End Simulation Exercise 4.2

4.8   Modeling State Machines

The examples that have been discussed so far were combinational and sequential circuits in isolation. Processes that model combinational circuits are sensitive to the inputs, being activated whenever an event occurs on an input signal. In contrast, sequential circuits retain information stored in internal devices such as flip-flops and latches. The values stored in these devices are referred to as the *state* of the circuit. The values of the output signals may now be computed as functions of the internal state and values of the input signals. The values of the state variables may also change as a function of the input signals, but are generally updated at discrete points in time determined by a periodic signal such as the clock. Provided with a finite number of storage elements, the number of unique states for the circuit is finite and such a circuit is referred to as a finite state machine.

Figure 4.20 shows a general model of a finite state machine. The circuit consists of a combinational component and a sequential component. The sequential component consists of memory elements, such as edge-triggered flip-flops, that record the state and are updated synchronously on the rising edge of the clock signal. The combinational component is made up of logic gates that compute two Boolean functions. The *output function* computes the values of the output signals. The *next-state function* computes the new values of the memory elements (i.e., the value of the next state). The state diagram is a commonly used method to capture the behavior of the state machine. From the state diagram in Figure 4.20, we know that if the machine is in state

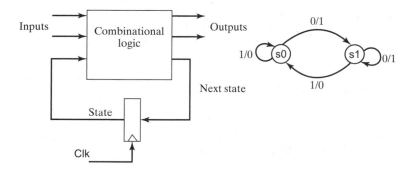

FIGURE 4.20   A behavioral model of a finite state machine

s0 and the input signal has a value of 0, then the machine transitions to state s1 while setting the value of the output signal to 1. The behavior of the state machine in state s1 can be similarly described. In fact, the behavior of larger state machines can be described in a similar manner—on a state-by-state basis. As we shall see shortly, such a description leads to a very natural VHDL description.

Figure 4.20 suggests a VHDL implementation using communicating concurrent processes. The combinational component can be implemented within one process. This process is sensitive to events on the input signals and changes in the state. Thus, if any of the input signals or the state variables change value, the process is executed to compute new values for the output signals and the new state variables. The sequential component can be implemented within a second process. This process is sensitive to the rising edge of the clock signal. When it is executed, the state variables are updated to reflect the value of the next state computed by the combinational component. The VHDL description of such a state machine is shown in Figure 4.21. The model is structured as two communicating processes, with the signals state and next_state used to communicate values between them. The structure of the process comb_process, representing the combinational component, is very intuitive. This process is constructed with a CASE statement. Each branch of the case represents one of the states and includes the output function and next-state function, as shown. The output value can be computed as a function of the current state and current input signal values. The next state can also be described as a function of the current state and current input signal values. The process clk_process updates the state variable on the rising edge of the clock. On reset, this process initializes the state machine to state state0. The process can be naturally extended to state machines composed of a larger number of states. If we can draw the state machine diagram in the form shown in Figure 4.20, then we can describe the behavior in each state as an additional branch of the case statement in process comb_process.

There are several interesting aspects to this model. First, note the use of enumerated types for the definition of a state. The model includes the definition of a new type referred to as statetype. This type can take on the values state0 and state1 and is referred to as an *enumerated type*, since we have enumerated all possible values that a signal of that type can take: in this case, exactly two distinct values, state0 and state1.

```vhdl
library IEEE;
use  IEEE.std_logic_1164.all;
entity state_machine is
port(reset, clk, x : in std_logic;
     z : out std_logic);
end entity state_machine;

architecture behavioral of state_machine is
type statetype is (state0, state1);
signal state, next_state : statetype := state0;
begin
comb_process: process (state, x) is
begin
 case state is                                -- depending upon the current state
 when state0 =>                               -- set output signals and next state
                 if x = '0' then
                 next_state <= state1;
                 z <= '1';
            else  next_state <= state0;
                 z <= '0';
            end if;
 when state1 =>
 if x = '1' then
                      next_state <= state0;
                      z <= '0';
            else  next_state <= state1;
                 z <= '1';
            end if;
 end case;
end process comb_process;

clk_process: process is
begin
wait until (rising_edge(clk)); -- wait until the rising edge
                 if reset = '1' then -- check for reset and initialize state
                 state <= statetype'left;
                 else   state <= next_state;
                 end if;
end process clk_process;
end architecture behavioral;
```

FIGURE 4.21   Implementation of the state machine in Figure 4.20

This enables much more readable and intuitive VHDL code. The case statement essentially describes the state machine diagram in Figure 4.20. In the clock process, note how the initial state is initialized on reset by using attributes discussed in Section 4.5. The clause statetype'left will return the value at the leftmost value of the enumeration of the possible values for statetype. Therefore, the initialization shown in the declaration is really not necessary. This is a common form of initialization and is also defined for other types in the language. The default initialization value for signals is signal_name'left. The use of such attributes during initialization provides a clean way of initializing signals without having to keep track of implementation-dependent values. A simulation of this state machine would produce a trace of the behavior, as shown in Figure 4.22. Note how the state labels appear in the trace, making it easier to read. Note also that the signal next_state is changing with the input signal, whereas the signal state is not. This is because state is updated only on the rising clock edge, while next_state changes whenever the input signal x changes.

We could just as easily have written the preceding example as a single process whose execution is initiated by the clock edge. In that case, the computation of the next state, the output signals, and the state transition are all synchronized by the clock. Alternatively, we could construct a model in which outputs are computed asynchronously with the computation of the next state. Such a model is shown in Figure 4.23. This model is constructed with three processes: one each for the output function, next-state function, and the state transition. Note how the model is constructed from the structure of the hardware: Concurrency in the circuit naturally appears as multiple concurrent processes in the VHDL model. State machines that compute the output signal values only as a function of the current state are referred to as Moore machines.

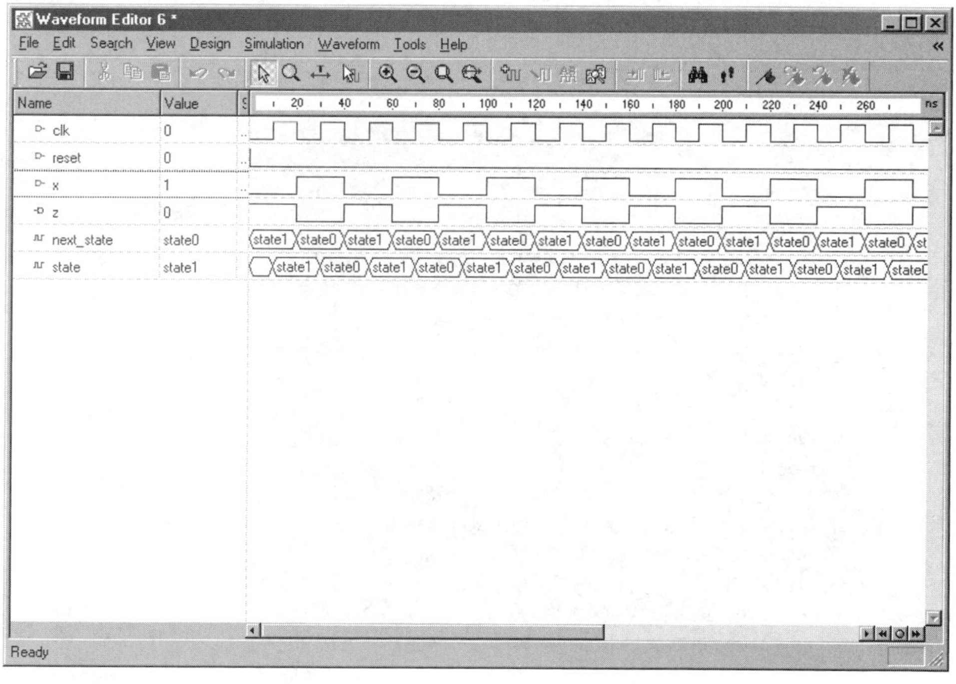

FIGURE 4.22 A trace of the operation of the state machine in Figure 4.21

```vhdl
library IEEE;
use IEEE.std_logic_1164.all;
entity state_machine is
port(reset, clk, x : in std_logic;
    z : out std_logic);
end entity state_machine;

architecture behavioral of state_machine is
type statetype is (state0, state1);
signal state, next_state :statetype :=state0;
begin
output_process: process (state, x) is
begin
 case state is            -- depending upon the current state
when state0 =>           -- set output signals and next state
                if x = '1' then z <= '0';
                else z <= '1';
                end if;
when state1 =>
                if x = '1' then z <= '0';
                else z <= '1';
                end if;
end case;
end process output_process;
next_state_process: process (state, x) is
begin
 case state is            -- depending upon the current state
when state0 =>           -- set output signals and next state
                if x = '1' then next_state <= state0;
                else next_state <= state1;
                end if;
when state1 =>
                if x = '1' then next_state <= state0;
                else next_state <= state1;
                end if;
end case;
end process next_state_process;

clk_process: process is
begin
wait until (rising_edge(clk));   -- wait until the rising edge
if reset = '1' then   state <= statetype'left;
else state <= next_state;
end if;
end process clk_process;
end architecture behavioral;
```

FIGURE 4.23   Alternative model for a finite state machine

State machines that compute the output values as a function of both the current state and the input values are Mealy machines. It is evident that the preceding approaches to constructing state machines enable the construction of both Moore and Mealy machines by appropriately coding the output function.

4.9   Constructing VHDL Models Using Processes

A prescription for writing VHDL models using processes can now be provided. The first step is the same as that described in Section 3.4 for constructing models using CSAs: We construct a fully annotated schematic of the system modeled. Figure 4.24 illustrates a template for constructing such VHDL behavioral models. One approach for translating the annotated schematic to a VHDL model described in the template of Figure 4.24 is as follows:

Construct_ Process_Model

1. At this point, I recommend using the IEEE 1164 value system. To do so, include the following two lines at the top of your model declaration:

> **library** IEEE;
> **use** IEEE.std_logic_1164.**all**;

You can declare single-bit signals to be of type **std_logic** and multibit quantities of type **std_logic_vector**. If you need arithmetic functions, you may consider adding the following packages:

> **use** IEEE.std_logic_1164.**all**;
> **use** IEEE. std_logic_unsigned.**all**;

The first package has the definitions of many arithmetic operators in objects of type **std_logic**. The second package is useful if we declare any objects of type unsigned, a type used for quantities such as memory addresses.

2. Select a name for the entity (**entity_name**) and write the entity description specifying each input and output signal port, its mode, and associated type.

3. Select a name for the architecture (**arch_name**) and write the architecture description. Place both the entity and architecture descriptions in the same file. (As we will see in Chapter 8, this is not necessary in general.)

   3.1 Within the architecture description, name and declare all of the internal signals used to connect the components. The architecture declaration states the type of each signal and may include initialization. The information you need is available from your fully annotated schematic.

   3.2 Each internal signal is driven by exactly one component. The computation of values on each internal signal can be described using a CSA or a process.

library library-name-1, library-name-2;

use library-name-1**.**package-name.all;

use library-name-2**.**package-name.all;

entity entity_name **is**

port(*input signals* : **in** *type*;

    output signals : **out** *type*);

end entity entity_name;

architecture arch_name **of** entity_name **is**

-- declare internal signals, you may have multiple signals of
-- different types

signal *internal signals* : *type* := initialization;

begin

```
label-1: process(-- sensitivity list --) is
--- declare variables to be used in the process
variable variable_names : type:= initialization;
       begin
-- process body
end process label-1;
```

First
Process

```
label-2: process is
--- declare variables to be used in the process
variable variable_names : type:= initialization;
       begin
wait until (-- predicate--);
-- sequential statements
wait until (-- predicate--);
-- sequential statements
end process label-2;
```

Second
Process

internal-signal *or* ports <= *simple, conditional, or selected CSA*

-- other processes or CSAs

end architecture arch_name;

FIGURE 4.24   A template for a VHDL model using CSAs and processes

Using CSAs

For each internal signal, select a concurrent signal assignment statement that expresses the value of this internal signal as a function of the signals that are inputs to that component. Use the value of the propagation delay through the component provided for that output signal.

Using a Process

Alternatively, if the computation of the signal values at the outputs of the component are too complex to represent with concurrent signal assignment statements, describe the behavior of the component with a process. One process can be used to compute the values of all of the output signals from that component.

3.2.1 Label the process. If you are using a sensitivity list, identify the signals that will activate the process and place them in the sensitivity list.

3.2.2 Declare the variables that the process uses.

3.2.3 Write the body of the process, computing the values of output signals and the relative time at which these output signals assume these values. These output signals may be internal to the architecture or may be port signals found in the entity description. If a sensitivity list is not used, specify wait statements at appropriate points in the process to specify when the process should suspend and when it should resume execution. It is an error to have both a sensitivity list and a wait statement within the process.

3.3 If there are signals that are driven by more than one source, the type of such signals must be a resolved type and must have a resolution function declared for their use. For our purposes, use the IEEE 1164 types std_logic for single-bit signals and std_logic_vector for bytes, words, or multibit quantities. These are resolved types. Make sure that you include the **library** clause and the **use** clause to include all of the definitions provided in the std_logic_1164 package.

3.4 If you are using any functions or type definitions provided by a third party, make sure that you have declared the appropriate library with the **library** clause and declared the use of this package via the presence of a **use** clause in your model.

The behavioral model will now appear structurally as shown in Figure 4.24, inclusive of both CSA and sequential statements.

Simulation Exercise 4.3: State Machines

Consider the state machine shown in Figure 4.25. This state machine has three inputs. The first is the reset signal, which initializes the machine to state 0. The second is a bit-serial input. The third is the clock input. The state machine recognizes the sequence 101 in the input sequence and sets the value of the output to 1. The value of the output remains at 1 until the state machine is reset.

Step 1. Write the VHDL model for this state machine. Use an enumerated type to represent the state (i.e., the type statetype as in Figure 4.23). Structure your state machine description as three processes:

Process 1: Write an output process that determines the value of the single-bit output on the basis of the current state and the value of the single-bit input.

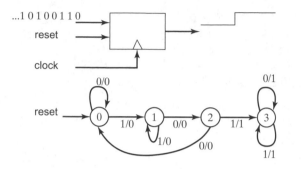

FIGURE 4.25   State machine for recognizing bit patterns

Process 2: Write a process that computes the next state on the basis of the value of the input signal and the current state.

Process 3: Write a clock process that updates the state on the rising edge of the clock signal. On a reset pulse the state machine is reset. Otherwise the state is modified to reflect the next state.

Step 2.   Within the simulator that you are using, structure the input stimulus as follows:

Apply a clock signal with a period of 20 ns.

Apply a reset signal that generates a single pulse of duration 30 ns.

Generate a random, bit-serial sequence with the pattern 101 embedded within the sequence.

Step 3.   Simulate the model long enough to detect the pattern in the input sequence.

Step 4.   Modify the model to recognize other patterns and repeat.

Step 5.   Modify the model so that, after the pattern is recognized, the state machine is reinitialized to state 0. Detection of the pattern now results in a pulse on the output signal.

End Simulation Exercise 4.3

4.10   Common Programming Errors

In this section we list some common programming errors made during the learning process.

4.10.1   Common Syntax Errors

- Do not forget the semicolon at the end of a statement.
- Remember, it is **elsif** and not **elseif**!

- It is an error to use **endif** instead of **end if**.
- It is an error to have a ";" after **then** in an **if-then-elsif** construct.
- Do not forget to leave a space between the number and the time base designation. For example, it should read 10 **ns** and not 10**ns**.
- Use underscores and not hyphens in label names. Thus, not half-adder, but half_adder.
- Expressions on the right-hand side of an assignment statement may have binary numbers expressed as x"00000000". When we use the IEEE 1164 types, even though std_logic_vector is a vector of bit signals, they are not the same type as x"00000000". Simulators may require you to perform a type conversion operation to convert the type of this binary number to std_logic_vector before you can make this assignment. The function to_stdlogicvector(x"00000000") is available in the package std_logic_1164 that is in the library IEEE. Check the documentation for the VHDL simulator you are using. Such type mismatches will be caught by the compiler.

4.10.2 Common Run-Time Errors

- It is not uncommon to use signals when you should use variables. Signals will be updated only after the next simulation cycle. Thus, you will find that values take effect later than you expected—for example, one clock cycle later.
- If you use more than one process to drive a signal, the value of the signal may be undefined unless you use resolved types and have specified a resolution function. Make sure that there is only one source (e.g., a process) for a signal, unless you mean to have shared signals. When using the std_logic_1164 package, use std_logic and std_logic_vector types. These are resolved types that provide an associated resolution function.
- A process should have a sensitivity list or a wait statement.
- A process cannot have both a sensitivity list and a wait statement.
- Remember that all processes will execute once when the simulation is started. This can sometimes cause unintended side effects if your processes are not explicitly controlled by wait statements.
- Suppose you have the following sequence of statements in a process:

```
wait until rising_edge(clk);
sig_a <= sig_x and sig_y;
sig_b <= sig_a;
```

When the second statement executes, sig_b will be assigned the value of sig_a from the previous execution of the process. The signal sig_b will not be assigned the value computed by the previous statement during the current execution of the process. To understand the semantics, consider the synchronous hardware implementation implied by the **wait** statement, where sig_a and sig_b are stored in flip-flops.

The output of the flip-flop holding the value of sig_a will be the input of the flip-flop corresponding to sig_b. Since there is a finite delay from inputs to outputs of the flip-flops, we see that this behavior is preserved in the semantics of the VHDL signal assignment statements.

4.11   Chapter Summary

This chapter has introduced models that use processes and has examined the use of sequential statements. This is a generalization of the behavioral models with concurrent signal assignment statements described in Chapter 3. The concepts introduced in the chapter include the following:

- Processes
- Sequential statements
  - if-then-else
  - case
  - loop
- Wait statements
- Attributes
- Communicating processes
- Modeling state machines
- Using both CSAs and processes within the same architecture description

Exercises

1. Construct a VHDL model of a parity generator for 7-bit words. The parity bit creates an even number of bits in the word with a value of 1. Do not prescribe propagation delays to any of the components. Simulate the model and check for functional correctness.

2. Explain why you cannot have both a sensitivity list and wait statements within a process.

3. Construct and test a model of a negative edge-triggered JK flip-flop.

4. Consider the construction of a register file with eight registers, where each register is 32 bits. Implement the model with two processes. One process reads the register file, while another writes the register file. You can implement the registers as signals declared within the architecture and therefore will remain visible to each process.

5. Implement a 32-bit ALU with support for the following operations: add, sub, and, or, and complement. The ALU should also produce an output signal that is asserted when the ALU output is 0. This signal may be used to implement branch instructions in a processor datapath.

6. Show an example of VHDL code that transforms an input periodic clock signal to an output signal at half the frequency.

7. Construct a VHDL model for generating four-phase nonoverlapping clock signals. Pick your own parameters for pulse width and pulse separation intervals.

8. Implement and test a 16-bit up–down counter.

9. Implement and test a VHDL model for the state machine for a traffic-light controller [7] shown in Figure 4.26.

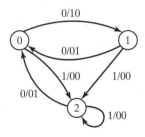

FIGURE 4.26   State machine for a traffic-light controller

10. Consider a variant of Simulation Exercise 4.3 in which we are interested in the occurrence of six 1's in the bit stream. After six 1's have been detected, the output remains asserted until the state machine is reset. Construct and test this model.

11. Consider the following code sequence:

```
proc2: process (x, y, z) is -- Process 2
begin
L1: sig_s1 <= (x and y) after 10 ns;
L2: sig_s2 <= (sig_s1 xor z) after 10 ns;
L3: res2 <= (sig_s1 nand sig_s2) after 10 ns;
end process;
```

At time 100, an event (0 to 1) on signal x activates the process. At this time, the values of signals sig_s1, sig_s2, y, and z are 0, 1, 1, and 0, respectively. What is the value of signal res2 scheduled for time 10 ns?

12. Write and test a VHDL model for checking equality of two 4-bit numbers.

13. Write and test a VHDL model for a BCD counter.

CHAPTER 5   Modeling Structure

Our model of a digital system remains that of an interconnected set of components. The preceding chapters described how we can specify the behavior of each component in VHDL. Informally, we can specify the behavior of each component as the set of output events that occur in response to input events. We can specify behavioral descriptions of a component via concurrent signal assignment statements. When this is infeasible due to the complexity of the event generation models, we can use one or more processes and sequential statements to specify the behavioral models of each component. A third approach to describing a system is simply in terms of the interconnection of its components. Rather than focusing on what each component does, we are concerned with simply describing how components are connected. Behavioral models of each component are already assumed to exist in the local working directory or in a library known to the VHDL simulator. Such a description is referred to as a *structural model*. Such models may describe only the structure of a system, without regard to the operation of individual components.

Several motivations drive the need for structural models. They enable the definition of a precise interface for the sharing of model components among developers within an organization. Structural models also facilitate the use of hierarchy and abstraction in modeling complex digital systems. As we will see at the end of this chapter, structural models are easily integrated with models that use processes and concurrent signal assignment statements, providing a powerful modeling approach for complex digital systems. The chapter discusses the basic principles governing the specification and construction of structural models in VHDL.

5.1   Describing Structure

A common means of conveying structural descriptions is through block diagrams. Components represented by blocks are interconnected by lines representing signals. Chapter 3 described models of a full adder with the use of concurrent signal assignment statements. These models provide a description of "what" the system does—namely, compute the value of signals and assign signals values at certain points in time relative to the current time. Instead of employing such a model, suppose we wish to describe the circuit as being constructed from two half adders under the assumption that we already understand the behavior of a half adder. Figure 5.1 shows such a design. In understanding how that design may be described in VHDL, imagine conveying this schematic over the telephone (no faxes!) to a friend who has no knowledge of full-adder circuits. You would like to have the friend correctly reproduce the schematic as you describe it. You can also think of describing this schematic across the table to someone without showing him or her the diagram or taking a pen to paper yourself. Do not use your fingers or hands to point, and use only verbal guidelines! When we think of conveying descriptions in these terms, we see that we need precise and unambiguous ways to describe structure.

We might convey such a description this way: First, we must describe the inputs and outputs to the full adder. This is not too difficult to convey verbally. We can easily state the type and mode of the input–output signals—for example, whether they are single-bit signals, input signals, or output signals. This information constitutes the corresponding entity description. Now imagine describing the interconnection of the components over the telephone. You would probably first list the components you need: two half adders and a two-input OR gate. Conveying such a list of components

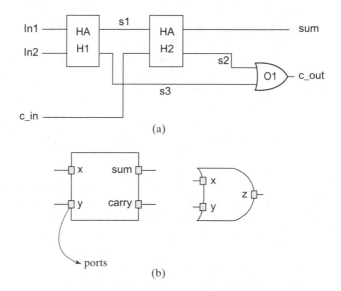

FIGURE 5.1   (a) Full-adder circuit (b) Interface description of the half-adder and OR-gate components

verbally is also not difficult, but now comes the tricky part: How do you describe the interconnection of these components unambiguously? In order to do so, you must first be able to distinguish between components of the same type. For example, in Figure 5.1(a), the half adders must be distinguished by assigning them unique labels, such as H1 and H2. The signals that connect these components are similarly labeled. For example, we may label them s1, s2, and s3. Additional annotations that we need are the labels for the ports in a half adder and the ports in an OR gate. Such detailed annotations are necessary so that we can refer to them unambiguously, such as the sum output port of half-adder H1 or the sum output port of half-adder H2. We have now completed the schematic annotation to the point where the circuit can be described in a manner that allows the person at the other end of the telephone to draw it correctly. For example, we can describe the interconnection between H1 and H2 by stating that the sum output of H1 is connected to signal s1 and the x input of H2 is connected to signal s1. Implicitly, we have stated that the sum output of H1 is connected to the x input of H2 by using signal s1. We have done this indirectly by describing connections between the ports and signal s1. There is an analogy with the physical process of constructing this circuit. If you were wiring the circuit on a protoboard in the laboratory, you would actually use signal wires to connect components. Once all of the components and their input and output ports are labeled, we see that we can describe this circuit in a manner that will allow the person to build it correctly.

On the basis of the preceding example, we can identify a number of features that a formal VHDL structural description might possess: (i) the ability to define the list of components, (ii) the definition of a set of signals that interconnect these components, (iii) the ability to uniquely label, and therefore distinguish, among multiple copies of the same component, and (iv) the ability to specify how signals are connected to ports. Figure 5.2 shows the VHDL syntax that realizes these features in an architecture description.

The component declaration includes a list of the components used and the input and output signals of each component, or, in VHDL terminology, the input and output ports of the component. The component declaration essentially states that this architecture will be using components named half_adder and or_2 gate, but so far has not stated how many of each type of component it will use. The declaration of the components is followed by the declaration of all of the signals that will be used to interconnect the components. These signals would correspond to the set of signal wires you might use in the laboratory. From a programming language point of view, we note similarities to the manner in which we construct Pascal or C programs. In a Pascal program, we declare the variables and data structures that we will use (e.g., arrays) and their types before we actually use them. In a VHDL structural model as shown in Figure 5.2, we declare all of the components and signals that we will use before we describe how they interconnect. Collectively, the component and signal declarations complete the declarative part of the **architecture** construct. That part is analogous to the parts list you would have if you were to build this circuit in the laboratory. One syntactic note: In VHDL'87, the syntax of the component declarations is a bit different, as identified in the Figure 5.2. Now all that remains is the process of actually "wiring" the components together. This feature is provided in the **architecture** body that follows the declarative part; it is delimited by the **begin** and **end** statements.

'87 vs. '93
☞

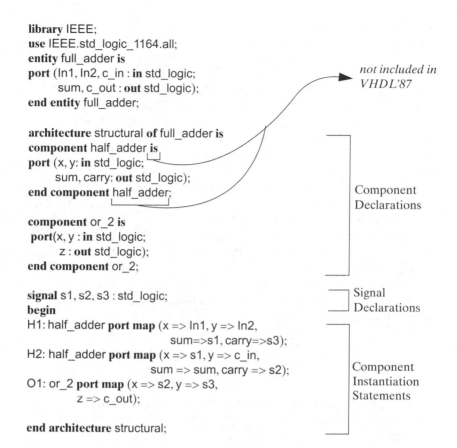

```
library IEEE;
use IEEE.std_logic_1164.all;
entity full_adder is
port (In1, In2, c_in : in std_logic;
      sum, c_out : out std_logic);
end entity full_adder;

architecture structural of full_adder is
component half_adder is
port (x, y: in std_logic;
      sum, carry: out std_logic);
end component half_adder;

component or_2 is
 port(x, y : in std_logic;
      z : out std_logic);
end component or_2;

signal s1, s2, s3 : std_logic;
begin
H1: half_adder port map (x => In1, y => In2,
                        sum=>s1, carry=>s3);
H2: half_adder port map (x => s1, y => c_in,
                        sum => sum, carry => s2);
O1: or_2 port map (x => s2, y => s3,
        z => c_out);

end architecture structural;
```

not included in VHDL'87

Component Declarations

Signal Declarations

Component Instantiation Statements

FIGURE 5.2 Structural model of a full adder

Consider the first statement in the architecture body. Let us go back to our analogy of wiring the circuit on a protoboard in the laboratory. We must first acquire the components, label them, and lay them out on the board. We must then connect each component to other components or circuit inputs by using the signal wires. Each line in the architecture body provides this information for each component. This is a component *instantiation statement*.

Recall that processes and assignment statements can be labeled. Components are similarly labeled. The first word, H1, is the label of a half-adder component. The **port map** ( ) construct that follows states how the input and output ports of H1 are connected to other signals and ports. The first argument of the port map construct simply states that the x input port of H1 is connected to the In1 port of the entity full_adder. The third argument states that the sum output port of H1 is connected to the s1 signal. The port map constructs in the remaining two component instantiation statements can be similarly interpreted. Note that both components have identical input port names. This is not a problem, since the name is associated with a specific component that is unambiguously labeled—namely, H1 and O1. When we instantiate a component, it is as if we were peering at the circuit through a keyhole and can see only one component at a time. We must completely describe the connections of all of the ports of each component before we can describe those of the next component. A close examination reveals that the interconnection of the schematic in Figure 5.1(a) is completely specified in the structural description of Figure 5.2.

Finally, one other important feature of this model should be noted. We assume that the behavioral models of each component are provided elsewhere; that is, there are entity–architecture pairs describing a half adder. The simulators and synthesis compilers must be able to find these models, which is typically achieved by matching the names provided in the models. For example, the structural model shown in Figure 5.2 states that a half-adder description, whose entity is labeled half_adder, is to be used. The simulator can find these models in the working directory, in a library, or in some directory contained in the search path that the user defines in the CAD tools. Note that there are no implications on the type of model used to describe the operation of the half adder. This behavioral model could comprise concurrent signal assignment statements, use processes and sequential statements, or itself be a structural model describing a half adder as the interconnection of gate-level behavioral models. We discuss such very useful hierarchies in greater detail later in the chapter.

Example: Structural Model of a State Machine

Consider the bit-serial adder shown in Figure 5.3. Two operands are applied serially, bit by bit, to the two inputs. On successive clock cycles, the combinational logic component

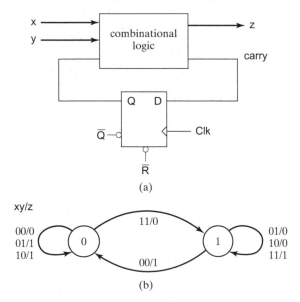

FIGURE 5.3   A bit-serial adder (a) Logic implementation (b) State diagram

computes the sum and carry values for each bit position. The D flip-flop stores the carry bit between additions of successive bits and is initialized to 0. As successive bits are added, the corresponding bits of the output value are produced serially on the output signal. The state machine diagram is shown below the circuit. As described in Section 4.8, we can develop this as a behavioral model of a state machine that implements the state diagram shown in the figure. In this example, we show the structural model of the state machine implementation. We have redrawn this circuit in the manner shown in Chapter 4 with a combinational logic component and a sequential logic component. Figure 5.4 shows the corresponding VHDL structural model. Note the use of the **open** clause for component labeled D1. The signal qbar at the output of the flip-flop is not used. The **port map**() clause in the component instantiation statement expresses this by setting the signal to **open**. The hardware analogy is that of an output pin on a chip that remains unconnected. Note also the first item in port map for C1, namely, x =>x. Initially, it may appear that this port map is ambiguous, since it may not be obvious which x is which. In fact, the parameter to the left of the => operator refers to a port on the component. The parameter to the right of the => operator refers to signal in the architecture or an entity port. The component has only one port named x, and the architecture and

```
library IEEE;
use IEEE.std_logic_1164.all;
entity serial_adder is
port (x, y, clk, reset : in std_logic;
      z : out std_logic);
end entity serial_adder;

architecture structural of serial_adder is
--
-- declare the components that we will be using
--
component comb is
 port (x, y, c_in : in std_logic;
       z, carry : out std_logic);
end component comb;
component dff is
port (clk, reset, d : in std_logic;
      q, qbar : out std_logic);
end component dff;
signal s1, s2 :std_logic;
begin
--
-- describe the component interconnection
--
C1: comb port map (x => x, y => y, c_in => s1, z =>z, carry => s2);
D1: dff port map(clk => clk, reset =>reset, d=> s2, q=>s1,
                 qbar => open);
end architecture structural;
```

FIGURE 5.4   VHDL structural model of the bit-serial adder

entity ports have only one signal named x. Thus, there is no ambiguity. Part of the rationale for permitting such syntactic arrangements follows from the need to reuse models developed by distinct users. We have no idea what other designers might name their ports. Thus, we cannot rely on the fact that all port names will be distinct. Imagine what would happen if we required everyone to use distinct port names. Every time there was a name conflict, it would take a significant amount of error-prone effort to rename ports and signals throughout large complex models. The syntax and scope rules for names used in the port maps ensure that the use of the same port names in distinct components does not pose a problem, thereby making it easier to share component models.

Example End: Structural Model of a State Machine

5.2   Constructing Structural VHDL Models

We are now ready to provide a prescription for constructing structural VHDL models. As with behavioral models, this simple methodology comprises two steps: (i) the drawing of an annotated schematic and (ii) the conversion to a VHDL model.

Construct_Structural_Schematic

1. Ensure that you have a behavioral or structural description of each component in the system being modeled. This means that you have a correctly working entity–architecture description of each component. Using the entity descriptions, create a block for each component with the input and output ports labeled.
2. Connect each port of each component to the port of another entity or to an input or output port of the system being modeled.
3. Label each component with a unique identifier: H1, U2, and so on.
4. Label each internal signal with a unique signal name and associate a type with this signal—for example, std_logic_vector. Make sure that the signals and ports which are connected are of the same type.
5. Label each system input port and output port and define its mode and type.

    This annotated schematic can be transcribed into a structural VHDL model. Figure 5.5 illustrates a template for writing structural models in VHDL. One approach to filling in this template is described in the following procedure, which relies on the availability of the annotated schematic.

Construct_Structural_Model

1. At this point, we recommend using the IEEE 1164 value system. To do so, include the following two lines at the top of your model declaration:

    ```
    library IEEE;
    use IEEE.std_logic_1164.all;
    ```

```
library library-name-1, library-name-2;
use library-name-1.package-name.all;
use library-name-2.package-name.all;

entity entity_name is
port( input signals : in type;
output signals : out type);
end entity entity_name;
architecture arch_name of entity_name is
-- declare components used

component component1_name is
port(input signals : in type;
        output signals : out type);
end component component1_name;

 component component2_name is
port( input signals : in type;
        output signals : out type);
end component component2_name;

        -- declare all signals used to connect the components

signal internal signals : type := initialization;
begin
        -- label each component and connect its ports to signals or entity ports

Label1: component1-name port map (port=> signal,.....);

Label2: component2-name port map (port => signal,.....);

end architecture arch_name;
```

FIGURE 5.5   Structural model template

Single-bit signals can be declared to be of type std_logic while multibit quantities can be declared to be of type std_logic_vector.

2. Select a name for the entity (entity_name) representing the system being modeled, and write the entity description. Specify each input and output signal port, its mode, and associated type.

3. Select a name for the architecture (arch_name), and write the architecture description as follows:

   3.1 Construct one component declaration for each unique component that the model will use. You can easily construct a component declaration from the component's entity description. The name used in the component declaration must be the same name used for the entity in the existing entity–architecture model of this component. That is how the simulator can find the model for the corresponding components when it is time to simulate the model.

3.2 Within the declarative region of the architecture description—before the **begin** statement—list the component declarations.

3.3 Following the component declarations, name and declare all of the internal signals used to connect the components. Your schematic shows these signal names. The declaration states the type of each signal and may also provide an initial value.

3.4 Now we can start writing the architecture body. For each block in your schematic, write a component instantiation statement using the **port map** () construct. The component label is derived from the schematic. The **port map** () construct will have as many entries as there are ports on the component. Each element of the port map construct has the form

port-signal => (internal signal or entity-port)

Each port of the component is connected to an internal signal or to a port of the top-level entity. Remember, the mode and type of the port of the component must match that of the internal signal or entity port that is connected to the component port.

4. Some signals may be driven from more than one source. Such signals are shared signals, and their type must be a resolved type. We can either define a new resolved type and its associated resolution function (as described in Chapter 6) or simply use the IEEE 1164 data types std_logic and std_logic_vector, which are resolved types.

It should be apparent that the process of generating a structural VHDL model from a schematic is a mechanical process. Thus, it is not surprising that modern CAD tools can generate such structural models automatically from design diagrams created with schematic capture tools available in current CAD environments.

Simulation Exercise 5.1: A Structural Model

The goal of this exercise is to introduce the reader to the construction, testing, and simulation of a simple structural model.

Step 1. Create a text file with the structural model of the full adder shown in Figure 5.2. Let us refer to this file as *full-adder.vhd*.

Step 2. Create a text file with the model of the half adder shown in Figure 3.2. Let us refer to this file as *half-adder.vhd*. Ensure that the entity name for the half adder in this file is the same as the name you have used for the half-adder component declaration in the full-adder structural model. Remember, the environment must have some way of being able to find and use the components that you need when you simulate the model of the full adder. Just as in the laboratory, component names are used for the purpose. In fact, the file names can be different as long as you consistently name architectures (arch_name), entities (entity_name), and components (component_name).

Some thought will reveal that such an approach follows intuition. Designs must be described in terms of lower level design units. File names are an artifact of the computer system we are using. When we compile an entity named E1 in a file called *homework1.vhd*, we will see that compiled unit names are based on the label E1 rather than *homework1.vhd*. This will enable compilation of higher level structural models to find relevant files on the basis of their component names and not the names of the files in which they are stored. Chapter 8 presents more on the programming mechanics.

Step 3. Create a text file with a model of a two-input OR gate. Let us refer to this file as *or2.vhd*. Use a gate delay of 5 ns. Again, make sure that the entity name is the same as the component name for the two-input OR gate model declared in the model of the full adder.

Step 4. Compile the files *or2.vhd*, *half-adder.vhd*, and *full-adder.vhd* in this order.

Step 5. Load the simulation model into the simulator.

Step 6. Open a trace window with the signals you would like to trace. Include internal signals which are signals that are not entity ports in the model.

Step 7. Generate a test case. Apply the stimulus corresponding to the test case to the inputs. Run the simulation for one time step. Examine the output to ensure that it is correct.

Step 8. Run the simulation for 50 ns.

Step 9. Check the behavior of the circuit and note the timing on the internal signals with respect to the component delays.

End Simulation Exercise 5.1

5.3   Hierarchy, Abstraction, and Accuracy

The structural model of the full adder shown in Figure 5.2 assumes the presence of models of the half adder. Although this model could be any one of the behavioral models described in Chapter 3 or Chapter 4, the model could also be a structural model itself, as shown in Figure 5.6. Thus, we have a hierarchy of models, graphically depicted in Figure 5.7. Each box in the figure denotes a VHDL model, an entity–architecture pair. The architecture component of each pair may in turn reference other entity–architecture pairs. At the lowest level of the hierarchy exist architectures composed of behavioral, rather than structural, models of the components.

A few interesting observations can be made about the model shown in Figure 5.6. We see that structural models simply describe interconnections; they do not describe

```
architecture structural of half_adder is
component xor2 is
 port (x, y : in std_logic;
        z : out std_logic);
end component xor2;
component and2 is
port (x, y : in std_logic;
        z : out std_logic);
end component and2;
begin
EX1: xor2 port map (x => x, y => y, z => sum);
OR1: and2 port map (x=> x, y=> y, z=> carry);
end architecture structural;
```

FIGURE 5.6 Structural model of a half adder

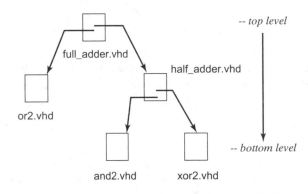

FIGURE 5.7 Hierarchy of models used in the full adder

any form of behavior. There are no descriptions of how output events are computed in response to input events. How can the simulation be performed? When the structural model is loaded into the simulator, the simulator creates a simulation model by replacing the components with their behavioral descriptions. If the description of a component is also a structural description (as is the case in the model of Figure 5.2, using the architecture of the half-adder model in Figure 5.6), then the process is repeated for structural models at each level of the hierarchy, until all components of the hierarchy have been replaced by behavioral descriptions. The levels of the hierarchy correspond to different levels of detail or *abstraction*. This process is referred to as *flattening* of the hierarchy. After the hierarchy has been flattened, we have a discrete event model that can be simulated. From this point of view, we see that structural models are a way of managing large, complex designs. A modern design may have several million to tens of millions of gates. It is infeasible to build a single, flat simulation model at the level of individual gates to test and evaluate the complete design. This may be due to the amount of simulation time required or the amount of memory required for such a detailed model. Thus, we may choose to approximate this gate-level behavior by constructing less accurate models. For example, we have seen that we may describe a state machine at the gate level or via a process-level description (as in Figure 4.20). The latter model is said to be at a higher level of abstraction. Managing large, complex designs requires the ability to work at multiple levels of abstraction, as indicated in the following examples:

1. We may have a library of VHDL models of distinct components, such as those derived from the manufacturer's component data book. These models have been developed, debugged, and tested. You can construct a model of a circuit simply by using these components. The only model you will have to write is a structural model. You also have to know the component entity description so that you correctly declare the component you are using. This is akin to using a library of mathematical functions in C or Pascal.

2. After three weeks, you might have a new and improved model of a half adder that you would like to use. You can test and debug this model in isolation. Then simply replace the old model with this new model. The need to recompile any dependent design entities is determined by the rules described in Chapter 8.

Finally, we note that the simulation time is directly affected by the level at which we construct simulation models. Consider the behavioral model of the half adder described in Chapter 4, Figure 4.3. Events on input signals produce events on output signals. In contrast, when we flatten the hierarchy of Figure 5.2, events on input signals will produce events on outputs of the gates within the half adders and eventually propagate to the output signals as events. The more detailed the model, the larger is the number of events we must expect the model to generate. The larger the number of events generated by the model, the greater is the simulation time. As a result, more accurate models will take a significantly longer time to simulate. Generally, the closer we are to making implementation decisions, the more accurate we wish the simulation to be, and hence more time is invested in simulation.

Since the full-adder model depends upon the existence of models for the half adder and two-input OR gate components, it follows that these models must be compiled before the model for the full adder is compiled. Once we have compiled all of the models, what happens if we make changes to only some design units? Must we recompile all of the design units each time we make a change to any one? If not, what dependencies between design units must we respect? These issues are discussed in Chapter 8.

Simulation Exercise 5.2: Construction of an 8-Bit ALU

The goal of this exercise is to introduce trade-offs in building models at different levels of abstraction and trading accuracy for simulation speed.

Step 1. Start with the model of a single-bit ALU as given in Simulation Exercise 3.2. This model is constructed with concurrent signal assignment statements.

Replace the model with one that replaces all of the concurrent signal assignment statements in the architecture body with sequential assignment statements and a single process. The process should be sensitive to events on input signals x, y, and c_in and should use variables to compute the value of the ALU output. The last statement in the process should be a signal assignment statement assigning the ALU output value to the signal result. Use a delay of 10 ns through the ALU.

Step 2. Analyze, simulate, and test the model, and ensure that all three operations (AND, OR, and ADD) operate correctly.

Step 3. Construct a VHDL structural model of a 4-bit ALU. Use the single-bit ALU as a building block. Use a ripple carry implementation to propagate the carry between single-bit ALUs.   Remember to compile the single-bit model before you compile the 4-bit model.

Step 4. Construct an 8-bit ALU using the 4-bit ALU as a building block. Use a ripple carry implementation to propagate the carry signal. Remember to compile the single-bit model before you analyze the 8-bit model.

Step 5. Based on your construction, what is the propagation delay though the 8-bit adder?

Step 6. Open a trace window with the signals you would like to trace. In this case, you will need to trace only the input and output signals to test the model.

Step 7. Generate a test case for each ALU instruction. Apply the stimulus for a test case to the inputs. Run the simulation for a period equal to at least the delay through the 8-bit adder. Examine the output values to ensure that the simulation is correct.

Step 8. Print the trace.

Step 9. Rewrite the 8-bit model as a behavioral model rather than a structural model. In this case, there is no hierarchy of components. Use a single process and the following hints:

   Step 9(a) Inputs, outputs, and internal variables are all now 8-bit vectors of type std_logic_vector (use the IEEE 1164 value system.)

   Step 9(b) Make use of variables to compute intermediate results.

   Step 9(c) Use the case statement to decode the opcode.

   Step 9(d) Do not forget to set the value of the output carry signal.

   Step 9(e) The propagation delay should be set to the delay through the hierarchical 8-bit model.

Step 10. Test the new model. In doing so, you should be able to use the same inputs you used for the hierarchical model.

Step 11. Generate a trace for the single-level model, demonstrating that the model functions correctly.

Step 12. Qualitatively compare the two models with respect to the difference in the number of events that occur in the flattened hierarchical model and the single-level model in response to a new set of inputs.

End Simulation Exercise 5.2

5.4   Generics

What if we would like to use the latest semiconductor technology that produces lower delays for our components? Suppose gate delays have now dropped from 5 ns to 1 ns. How do we keep our models current? We can run through them with a text editor and change all of the gate delays. This is clearly undesirable. The right answer is to be able to construct parameterized models rather than have hard-coded delays. The actual value of the gate delay is determined at simulation time by the value that is provided to the model. Having parameterized models makes it possible to construct standardized libraries of models that can be shared. The issue to be addressed in creating parameterized models is the accommodation of hierarchical models composed of many components organized in multiple levels whose relative delays are not independent, but rather maintain a fixed relationship. The VHDL language provides the ability to construct such parameterized models using the concept of *generics*.

Figure 5.8 illustrates a parameterized behavioral model of a two-input exclusive-OR gate. The propagation delay in this model is parameterized by the constant gate_delay. The default (or initialized value) value of gate_delay is set to 2 ns. This is the value of delay that will be used in simulation models such as that shown in Figure 5.6, unless a different value is specified. A new value of gate_delay can be specified at the time the model is used, as shown in Figure 5.9. This version of the half adder will make use of exclusive-OR gates that exhibit a propagation delay of 6 ns through the use of the **generic map**() construct. Note the absence of the ';' after the **generic map**() construct!

```
library IEEE;
use IEEE.std_logic_1164.all;

entity xor2 is
generic (gate_delay : Time:= 2 ns);
port (In1, In2 : in std_logic;
     z : out std_logic);
end entity xor2;

architecture behavioral of xor2 is
begin
z <= (In1 xor In2) after gate_delay;
end architecture behavioral;
```

FIGURE 5.8   An example of the use of generics

```
architecture generic_delay of half_adder is
component xor2 is
generic (gate_delay: Time); -- new value may be specified here instead
port (x, y : in std_logic;      -- of using a generic map() construct
     z : out std_logic);
end component xor2;
component and2 is
generic (gate_delay: Time);
port (x, y : in std_logic;
     z : out std_logic);
end component and2 ;
begin
EX1: xor2 generic map (gate_delay => 6 ns)
          port map(x => x, y => y, z => sum);
A1: and2 generic map (gate_delay => 3 ns)
          port map(x=> x, y=> y, z=> carry);
end architecture generic_delay;
```

FIGURE 5.9   Use of generics in constructing parameterized models

The xor2 model is now quite general. Rather than manually editing and updating the delay values in the VHDL text, we can specify the value we must use when the component is instantiated. Considering the thousands of digital system components that are available, manually modifying each model when we wish to change its attributes can be quite tedious, inefficient, and error prone.

5.4.1   Specifying Generic Values

The example in Figure 5.9 illustrates how we can specify the value of a generic constant by using the **generic map**() construct when the component is instantiated. Alternatively, we can specify the value when the component is declared. For example, in Figure 5.8 the component declaration of xor2 includes a declaration of the generic parameters. We can modify this statement to appear as follows:

generic (gate_delay: **Time**:= 6 **ns**);

Therefore, we observe that, within a structural model, there are at least two ways in which the values of generic constants of lower-level components can be specified: (i) in the component declaration and (ii) in the component instantiation statement, using the **generic map**() construct. If both are specified, then the value provided by the **generic map**() takes precedence. If neither is specified, then the default value defined in the model is used.

The values of these generics can be passed down through multiple levels of the hierarchy. For example, suppose that the full-adder model shown in Figure 5.2 defines the value of gate_delay for all lower level modules. In this case, the half-adder module may be modified to appear as shown in Figure 5.10.

```
library IEEE;
use IEEE.std_logic_1164.all;

entity half_adder is
generic (gate_delay:Time:= 3 ns);
port (a, b : in std_logic;
      sum, carry : out std_logic);
end entity half_adder;

architecture generic_delay2 of half_adder is
component xor2 is
generic (gate_delay: Time);
 port (x,y : in std_logic;
       z : out std_logic);
end component xor2;

component and2 is
generic (gate_delay: Time);
port (x, y : in std_logic;
      z : out std_logic);
end component and2;

begin
EX1: xor2 generic map (gate_delay => gate_delay)
         port map(x => x, y => y, z => sum);
A1: and2 generic map (gate_delay => gate_delay)
         port map(x=> x, y=> y, z=> carry);
end architecture generic_delay2;
```

FIGURE 5.10   Passing values of generics through multiple levels of the hierarchy

Within the half adder, the default gate delay is set to 3 ns. This is the value of gate_delay that is normally passed into the lower level models for xor2 and and2. However, if the full-adder description uses **generic maps** to provide a new value of the gate delay to the half-adder models that are instantiated in the structural description shown in Figure 5.2, then these values take precedence and will flow down to the gate-level models. This process is depicted graphically in Figure 5.11. From the figure, it is evident that, by changing the value of gate_delay in the full-adder model, the gate-level VHDL models for xor2 and and2 will utilize this value of the gate delay in the simulation. Although the models are written with default gate-delay values of 2 ns, the new value overrides this default value.

5.4.2   Some Rules about Using Generics

The terminology is quite appropriate: The use of generics enables us to write *generic* models whose behavior in a particular simulation is determined by the value of the

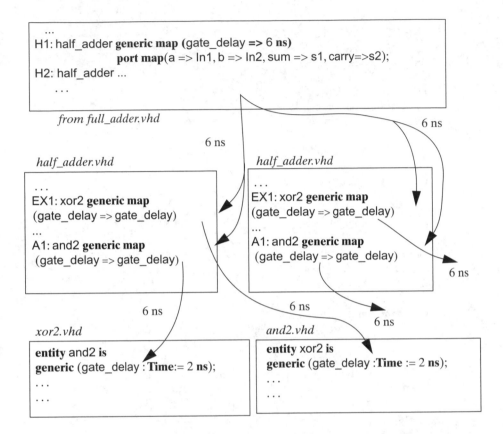

FIGURE 5.11 Parameter passing through the hierarchy using generics

generic parameters. Generics appear very much like ports. They are a part of the interface specification of the component. However, unlike ports, they do not have a physical interpretation. They are more a means of conveying information through the design hierarchy, thereby enabling component designs to be parameterized. Generics are constant objects. Therefore, they cannot be written, but only read. The values of generic parameters must be computable at the time the simulator is loaded with the VHDL model. Therefore, we may include expressions in the value of a generic parameter. However, the value of this expression must be computable at the time the simulator is loaded. Finally, we must be careful about the precedence of the values of generic objects, as described in the preceding sections.

The use of generics is not limited to the specification of delays. In fact, the most powerful uses of generics deal with physical attributes other than delays. The next few examples further illustrate this power of constructing parameterized models.

Example: N-Input OR Gate

In one class of generic gate-level models, the number of inputs can be parameterized. Therefore, we can have just one VHDL model of an N-input gate. We can produce a 2-, 3-, or 6-input gate model by setting the value of a generic parameter. Figure 5.12 illustrates this technique. When this OR-gate model is used in a VHDL model, the generic parameter n must be mapped to the required number of inputs, using the **generic map ()** construct in a higher level structural model. We see how such models are constructed. The VHDL code uses a loop to perform the gate operation in a bit-serial fashion. The generic parameter defines the loop index and therefore the number of bits at the gate input. The bit-serial nature of the model decouples the computation from a fixed bit width.

If we were using several OR gates in a structural model, we would have one instantiation statement for each model. Each instantiation statement could include the following statement for different values of n: **generic map** (n=>3). In this statement, the value 3 would be replaced by the number of inputs for that particular gate. Note that there is only one VHDL model, but we can use it to instantiate multiple gates of

```vhdl
library IEEE;
use IEEE.std_logic_1164.all;
entity generic_or is
generic (n: positive:=2);
port (in1 : in std_logic_vector ((n-1) downto 0);
    z : out std_logic);
end entity generic_or;

architecture behavioral of generic_or is
begin
process (in1) is
variable sum : std_logic:= '0';
begin
sum := '0'; -- on an input signal transition sum must be reset to 0
for i in 0 to (n-1) loop
sum := sum or in1(i);
end loop;
z <= sum;
end process;
end architecture behavioral;
```

FIGURE 5.12   An example of a parameterized gate model

different input widths. The conventional programming language analogy would be that of using functions for which the arguments determine the computation performed by the function.

Example End: N-Input OR Gate

Example: N-Bit Register

Another example of the use of generics is a parameterized model of an N-bit register. We can use generics to configure the model in a specific instance to be of a fixed number of bits. Let us consider a register made up of D flip-flops with asynchronous reset and load enable signals. The model is shown in Figure 5.13. The operation of the register (process) is sensitive to the occurrence of an event on the reset or clk signals. The size of the register is determined when this model is instantiated by a higher level model. The default value is a 2-bit register. Note the predefined type we have used for the generic parameter, namely, **positive**. This type is defined in VHDL to be a subtype of an integer with a range starting from 1 rather than 0. Such a definition forces the generic parameter to have a value of at least 1, and it cannot be less than 1. There can be no registers with 0 bits, which certainly makes sense. In general, when

```
library IEEE;
use IEEE.std_logic_1164.all;
entity generic_reg is
generic (n: positive:=2);
port ( clk, reset, enable : in std_logic;
      d : in std_logic_vector (n-1 downto 0);
      q : out std_logic_vector (n-1 downto 0));
end entity generic_reg;

architecture behavioral of generic_reg is
begin
reg_process: process (clk, reset) is
begin
 if reset = '1' then
    q <= (others => '0');
  elsif (rising_edge(clk)) then
     if enable = '1' then
       q <= d;
    end if;
   end if;
 end process reg_process;
end architecture behavioral;
```

FIGURE 5.13   An example of a parameterized model of an N-bit register

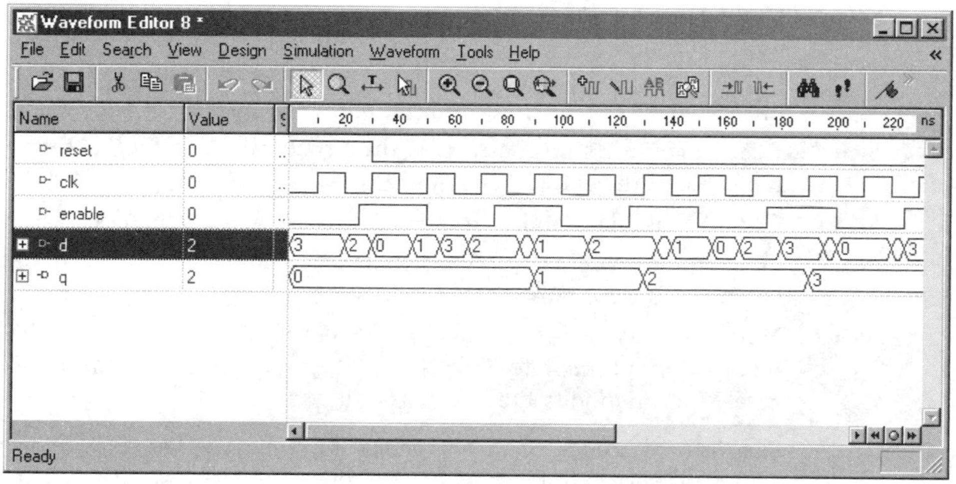

FIGURE 5.14   Trace of the operation of a 2-bit register

we are parameterizing physical quantities, it is usually the case that a value of 0 or negative values do not make sense. In these instances, it is natural to use the type positive.

Note how the value of q is set with the use of the "**others**" construct. Since q is a vector of bits, this statement provides a concise approach to specifying the values of all the bits in a vector when they are equal. A trace of the operation of this 2-bit register in the presence of various waveforms on the input signals is also shown in Figure 5.14.

Example End: N-Bit Register

Simulation Exercise 5.3: Use of Generics

This exercise illustrates the utility of the use of generics for parameter passing in structural models.

Step 1.   Start with the model of a single-bit ALU that is written using CSAs from Simulation Exercise 3.2. Modify this model to include a generic parameter to specify the gate delay. Set the default gate delay to 3 ns. Use 2 ns for the delay through the multiplexor.

Step 2.   Compile, simulate, and test the model and ensure that all three operations (AND, OR, and ADD) are correctly computed using the default value of the gate delay.

Step 3.   Construct a VHDL structural model of a 2-bit ALU. Use the single-bit ALU as a building block. Use a ripple carry implementation to propagate the carry between single-bit ALUs. Remember to analyze (compile) the single-bit model before you analyze (compile) the 2-bit model.

Step 4.   Use the **generic map** construct in the structural model of the 2-bit ALU to set the value of the gate delay to 4 ns.

Step 5.   Compile, simulate, and verify the functionality of this model.

Step 6.   Open a trace window with the signals you would like to trace. In this case, you will need to trace only the input and output signals to test the model.

Step 7.   Note how easy it is to modify the values of the gate delay at the top level, recompile, and simulate. Now change the model to use a generic n-input AND gate. You can modify the model as follows:

    Step 7(a)   Create a model of a generic n-input AND gate to follow the model shown in Figure 5.12. Compile, simulate, and test this model in isolation. Make sure that it is in the same working directory as the rest of your model.

    Step 7(b)   Modify the single-bit ALU model to declare the generic AND model as a component and specify a default value of 3 for the number of inputs to the AND gate.

    Step 7(c)   In the model of the single-bit ALU, replace the CSA describing the operation of the AND gate with a component instantiation statement for the generic AND model. This instantiation statement should also include a **generic map** statement providing the number of gate inputs as a parameter.

    Step 7(d)   Since the default value of the number of inputs to the AND gate is 3, you must pass parameters correctly in order for this model to function.

Step 8.   Compile and test your model. Trace the input and output signals to determine that the model functions correctly.

Step 9.   Experiment with other possibilities. For example, you can use a generic model of an OR gate as well. This model is shown in Figure 5.12.

End Simulation Exercise 5.3

It should be apparent now that, from the point of view of writing models, a component instantiation statement is on par with a concurrent signal assignment statement or a process construct. The body of an architecture may be composed of all three types of statements, and they individually model concurrent activities. The declaration part of the **architecture** construct may now include declarations of signals as well as components. Thus, we see that the term *structural models* does not imply that we can describe the digital system or circuit only as an interconnected set of components. Instead, we can mix component instantiation statements with signal assignment statements and processes. Figure 5.15 shows a template for a general VHDL model.

```
library library-name-1, library-name-2;
use library-name-1.package-name.all;
use library-name-2.package-name.all;

entity entity_name is
port( input signals : in type;
output signals : out type);
end entity entity_name;

architecture arch_name of entity_name is
-- declare components used
component component1_name is
port( input signals : in type;
output signals : out type);
end component component1_name;

component component2_name is
port( input signals : in type;
output signals : out type);
end component component2_name;
-- declare all signals used to connect the components
signal internal signals : type := initialization;

begin
-- label each component and connect its ports to signals or other ports

Label1: component1-name port map (port=> signal,.....);

Label2: component2-name port map (port => signal,.....);

        -- we can include behavioral modeling statements that execute
        -- concurrently

concurrent signal assignment statement-1;
concurrent signal assignment statement-2;

process-1: ..

process-2:..

end architecture arch_name;
```

FIGURE 5.15   Component instantiation in behavioral models

5.5   The Generate Statement

In the preceding sections, the construction of structural models was based on instantiating the components in a design. Each component instantiation statement described
how the component inputs and outputs were connected to other signals in the design.
As a result, if we create designs with a large number of components, we write a large
number of component instantiation statements. Apart from the obvious time invest-
ment, opportunities for errors increase.

Suppose the design we are describing has a very regular structure. It would be
highly economical to be able to describe this structure and let the language or CAD
tools worry about the detailed list of component instantiation statements. For example, we can easily describe the construction of a 32-bit register from D flip-flops more
compactly than having to write 32 component instantiation statements. The VHDL
language provides such descriptive power in the form of the **generate** statement. The
choice of terminology is intentional and descriptive. Let us start with the simple example of a 32-bit register. Then we will proceed to a more complex example.

Example: N-Bit Register

We assume the availability of a model of a 1-bit D flip-flop named dff. Using this component, we wish to construct a 4-bit register. A straightforward model constructed
via the techniques described earlier in this chapter is shown in Figure 5.16. The model

```
library IEEE;
use IEEE.std_logic_1164.all;

entity dregister is
port (din : in std_logic_vector(3 downto 0);
      qout : out std_logic_vector(3 downto 0);
      clk : in std_logic);
end entity dregisters;

architecture behavioral of dregister is
component dff is
port (d, clk : in std_logic;
      q : out std_logic);
end component dff;

begin
      bit0: dff port map( d=>din(0), q=>qout(0), clk=>clk);
      bit1: dff port map( d=>din(1), q=>qout(1), clk=>clk);
      bit2: dff port map( d=>din(2), q=>qout(2), clk=>clk);
      bit3: dff port map( d=>din(3), q=>qout(3), clk=>clk);
end architecture dregister;
```

FIGURE 5.16   Creating an N-bit register using the component instantiation statements

```
library IEEE;
use IEEE.std_logic_1164.all;

entity dregister is
generic (width : natural:=16); -- default width of the register is 16
port (din : in std_logic_vector(width-1 downto 0);
      qout : out std_logic_vector(width-1 downto 0);
      clk : in std_logic);
end entity dregisters;

architecture behavioral of dregister is
component dff is
port (d, clk : in std_logic;
      q : out std_logic);
end component dff;

begin
    dreg: for i in d'range generate
    reg: dff port map( d=>din(i), q=>qout(i), clk=>clk);
    end generate;
end architecture dregister;
```

FIGURE 5.17   The model in Figure 5.16 rewritten using the generate statement

is composed of four component instantiation statements—one for each flip-flop that is used to construct the register.

The **generate** statement can be used to describe such a model more compactly. For example, note the arguments for the signals din and qout in the architecture body. In each statement, the value increases by one. In conventional programming languages, we do not add two arrays by having one assignment for each pair of elements. Rather, we write a loop. The **generate** statement can be used to perform the same function. Figure 5.17 shows a rewritten example from Figure 5.16, using the **generate** statement.

The structure is similar to that of a loop. In addition, the model uses a generic parameter to set the bit width, resulting in a very general, configurable register model. Each iteration through the loop effectively creates one of the component instantiation statements in Figure 5.16.

Example End: N-Bit Register

The majority of structures one typically encounters will not be as regular as registers. Rather, a large portion of the design may be regular, with a few components that warrant special treatment. Moreover, the individual components are not likely to be independent of the other components, as individual register bits are. There may be dependencies among components in the form of input/output signals. For example, consider the case of an 8-bit adder. For a ripple carry adder, full adders used in adjacent bit positions are connected by the carry signals between them. Models for such circuits employ the **generate** statement for portions of the design and can rely on general component instantiation statements for the remaining components. The next example will illustrate this point.

Example: N-Bit ALU

This example illustrates the use of the generate statement for the construction of a multibit arithmetic logic unit (ALU). Let us assume that we have a model of a single-bit ALU described in Simulation Exercise 3.2 and shown in Figure 3.10, and the name of this model is one_bit. This model can perform single-bit addition, logical AND, and logical OR on a pair of bits. The actual operation is selected by the opcode input signal. Our goal is to construct a multibit model using the generate statement. We arrive at the VHDL model as follows:

The ALU model will use ripple carry for addition. In this case, the ALUs at bit positions 0 and 7 differ in their external interconnects from the ALUs in bit positions 1 through 6. Consider an ALU in any of the positions 1 through 6. Any such ALU, i, will receive a carry input from ALU i-1 and will generate a carry signal for ALU i+1. The inputs will be drawn from entity input bit i, and the outputs will produce values for entity output bit i. The ALU at bit position 0 differs in that the carry input will be from the ALU entity carry input. The ALU in bit position 7 differs similarly in that the carry that is generated must be connected to the carry output of the ALU entity. Thus, we can use a generate statement to instantiate ALUs in bit positions 1 through 6 while instantiating the ALUs in bit positions 0 and 7 separately.

Note the need to provide separate **generic map** statements. Finally, to connect the individual carry signals we declare a local array of signals, carry_vector. From the code, it is clear that declaring these signals as an array is necessary to be able to write the **generate** statement.

Example End: N-Bit ALU

You will notice that all of the statements in the example shown in Figure 5.18 are labeled. The language requires all component instantiation statements, as well as all **generate** statements, to be labeled. Note that the labels by themselves have no active role. Thus, the label a2to6 could just as easily have been GTech. However, in general, the use of meaningful labels and names is, of course, encouraged. Further, as you might expect, generate statements can be nested.

```
library IEEE;
use IEEE.std_logic_1164.all;

entity multi_bit_generate is
generic(gate_delay:time:= 1 ns;
        width:natural:=8); -- the default is an 8-bit ALU
port( in1 : in std_logic_vector(width-1 downto 0);
     in2 : in std_logic_vector(width-1 downto 0);
     result : out std_logic_vector(width-1 downto 0);
     opcode : in std_logic_vector(1 downto 0);
     cin : in std_logic;
     cout : out std_logic);
end entity multi_bit_generate;

architecture behavioral of multi_bit_generate is
component one_bit is -- declare the single-bit ALU
generic (gate_delay:time);
port (in1, in2, cin : in std_logic;
     result, cout : out std_logic;
     opcode: in std_logic_vector (1 downto 0));
end component one_bit;

signal carry_vector: std_logic_vector(6 downto 0); -- the set of signals for the
                                                    -- ripple carry
begin
a0: one_bit generic map (gate_delay) -- instantiate ALU for bit position 0
port map (in1=>in1(0), in2=>in2(0), result=>result(0), =>cin, opcode=>opcode,
cout=>carry_vector(0));

a2to6: for i in 1 to width-2 generate -- generate instantiations for bit positions 2–6
a1: one_bit generic map (gate_delay)
port map(in1=>in1(i), in2=> in2(i), cin=>carry_vector(i-1),
result=>result(i), cout=>carry_vector(i),opcode=>opcode);
end generate;

a7: one_bit generic map (gate_delay) -- instantiate ALU for bit position 7
port map (in1=>in1(width-1), in2=>in2(width-1), result=> result(width-1),
cin=>carry_vector(width-2), opcode=>opcode, cout=>cout);
end architecture behavioral;
```

FIGURE 5.18 Generating a multibit ALU

From the preceding examples, we can uncover a few general principles to help guide us in the use of the **generate** statement when we have arrays of components that we wish to instantiate:

- Structure the array of components indexed by i, j, and so on.

- Declare, as necessary, signal vectors local to the architecture to interconnect the components.

- Determine which range of components can be encapsulated within a generate statement.

- Write one component instantiation statement for each of the remaining components.

In trying to determine how to index the components and signal vectors correctly, it is useful to think of sequential program loops whose loop bodies process array data structures. It is necessary to exercise care, since the compiler does not necessarily report unconnected ports on a component. The resulting design can easily lead to erroneous results, because these unconnected signals are simply modeled as unconnected signals in the target design. This can lead to the propagation of undefined signal values through the design, as well as the computation of incorrect signal values.

5.6   Configurations

We have seen that there are many different ways in which to model the operation of a digital circuit utilizing behavioral and structural models. Approaches to the construction of behavioral models may differ in their use of concurrent or sequential statements. Structural models may employ multiple levels of abstraction, and each component within a structural model may, in turn, be described as a behavioral or structural model. Consider the structural model of the full adder shown in Figure 5.2. Assume that two alternate architectures exist for the half-adder components. Design–1 is a behavioral model, as specified in Figure 3.2. Design–2 is a structural model, as specified in Figure 5.6. When the full-adder model shown in Figure 5.2 is compiled and simulated, which architecture for the half adder should be used? You would like to configure your simulation model to be able to use one or the other. The VHDL language provides *configurations* for explicitly associating an architecture description with each component in a structural model. This process of association is referred to as *binding* an instance of the component (in this example, the half adder) to an architecture. In the absence of any programmer-supplied configuration information, *default* binding rules apply.

Figure 5.19 graphically illustrates the notion of configurations. A design will have a single entity that describes how the design interfaces to its environment. The implementation is captured in the corresponding architecture, which may instantiate several components. For each of these components there are multiple possible implementations, each captured in an entity–architecture pair. In fact, all of these alternative implementations will have identical entity descriptions, but distinct architecture descriptions. To construct a simulation, we use configurations to specify a specific entity–architecture for each component that is instantiated in the top-level model.

Rather than jump right into the syntax and semantics of configurations, we are better served if we can at first generate an intuition about the nature of configurations. What better way to do this than by examples?

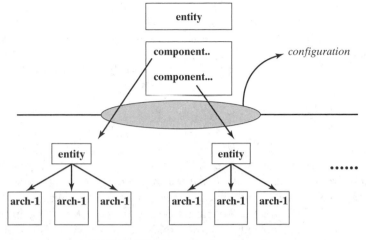

FIGURE 5.19   The role of configurations

Example: Component Binding

As an example of binding architectures to components, consider the structural model of a state machine for bit-serial addition shown in Figure 5.4. The component C1 implements the combinational logic portion of the state machine. There may be alternative implementations of the gate-level design of C1 corresponding to alternative designs for high speed, low power, parts from different vendors, or even simply a behavioral model written for simulation. One of these alternative models must be *bound* to the component C1 for simulation. The configuration construct in VHDL specifies one, and only one, such binding.

Note that we are concerned only with binding the combinational logic component with an architecture and are not concerned with configuring the entity description of C1. This is because the interface does not change, and therefore all alternative architectures for C1 will share an identical entity description. In fact, the entity description is identical to the component declaration for C1. It is the implementation of C1 that may change, and this is captured in an architecture for C1.

Example End: Component Binding

Notice how easy it is to analyze different implementations. We simply change the configuration, compile, and simulate. Configurations also make it easy to share designs. When newer component models become available, we can bind the new architecture to the component by editing the configuration information, then compile and simulate. We do not have to modify the VHDL structural model.

There are two ways in which configuration information can be provided: *configuration specification* and *configuration declaration*. But before discussing these, let us state the default binding rules that have been in effect for the examples presented in this chapter and explain how the VHDL tools find the architectures for the components in a structural model when no configuration information is provided.

5.6.1 Default Binding Rules

The structural model is simply a description of a schematic. Revisiting our analogy with the construction of a circuit on a protoboard, we now have to build the operational circuit. Assume that we can realize each component with a single chip from one of many vendors. We can think of the configurations as describing the chips we need to obtain and place on the board: one for each component. What if no configuration information were provided? How would we know what chip to use for each component and where to get them? Clearly, we expect some rules for doing so. For example, we might first look around the lab bench for chips with the same names as the components. If we could not find them, we might then look in the "usual" cabinet where all chips are stored, again seeking chips with the same names as the component names. Such rules are analogous to the default binding rules in VHDL.

If no configuration information is provided, as in the preceding examples, then we can find a default architecture as follows: If the entity name is the same as the component name, then this entity is bound to the component. For example, for the structural model in Figure 5.4, no configuration information is provided. The language rules enable a search for entities with the same names, in this case, comb and dff. These may be found in the working directory, in which case the architectures associated with these entities are used. However, what if there are multiple architectures for the entity, such as for comb as shown in Figure 5.20? In this case, we use the last compiled architecture for entity comb. Now the environment can construct a complete simulation model. Using the default rules, we see that, by using component names that are the same as the entity names, we can avoid writing configurations. This is what we have managed to do up to this point in this chapter. Using the same names for components and their entities also improves the readability of the code.

However, if no such entity with the same name is visible to the VHDL environment, then the binding is said to be deferred; that is, no binding takes place now, but information will be forthcoming later, as described in the next sections. This is akin to going ahead with wiring the rest of the circuit and hoping that your partner comes up with the right chips before you are ready to run the experiment!

5.6.2 Configuration Specification

Configuration specifications are used in the architecture body to identify the relationships between the components and the entity–architecture pairs used to model each component. Continuing with our laboratory analogy, consider how we might specify

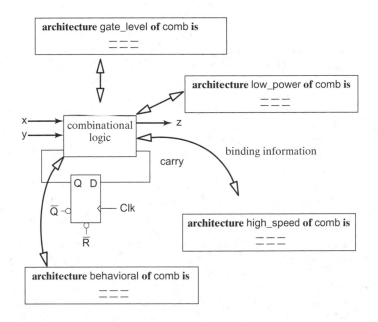

FIGURE 5.20   Alternative architectures for binding the combinational logic component of a state machine

the chips to be used for a component that we have declared. We might specify the chip name and location, for example, in the box labeled half adders in the grey cabinet. How can we similarly define the exact location of an entity–architecture pair? We can do so by naming the design library within which it is located and the name of the design unit within which such pairs are stored. The syntax is shown in Figure 5.21.

For the first half-adder component, we state that the entity description can be found in the library WORK. In most systems, libraries are implemented as directories. The library WORK is a special library and is usually the default working directory. Since an entity can have multiple architectures, we can specify the name of the architecture to be used for this entity within parentheses. Notice that, for the second half adder, the configuration specification uses a different architecture. If only an entity, and no architecture, is specified, then the last compiled architecture for that entity is used. All components of the same type need not use the same architecture or even the same entity! The configuration of the two-input OR gate deserves special attention. In this case, we have chosen to use an entity with a name different from the component. Therefore, we have to provide additional information. For example, if you were simply given a new chip named lpo2, you would need to know which pins corresponded to inputs and which corresponded to outputs before you could use the chip in place of the or_2 component. This information is provided as part of the configuration statement, which the figure shows via a **port map** () construct. We are using an entity named lpo2 that has been compiled into a library named POWER. The corresponding architecture that we will use is named behavioral and can be found in the same library. Similar arguments hold for any generic parameters.

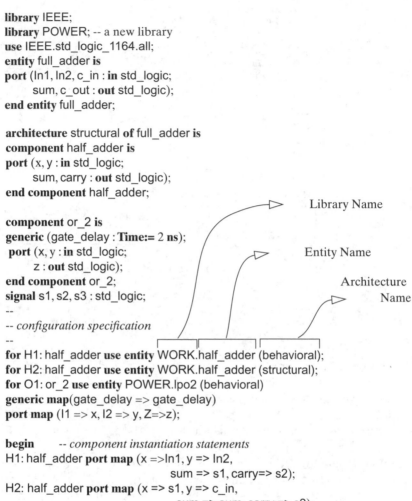

```
library IEEE;
library POWER; -- a new library
use IEEE.std_logic_1164.all;
entity full_adder is
port (In1, In2, c_in : in std_logic;
      sum, c_out : out std_logic);
end entity full_adder;

architecture structural of full_adder is
component half_adder is
port (x, y : in std_logic;
      sum, carry : out std_logic);
end component half_adder;

component or_2 is
generic (gate_delay : Time:= 2 ns);
 port (x, y : in std_logic;
      z : out std_logic);
end component or_2;
signal s1, s2, s3 : std_logic;
--
-- configuration specification
--
for H1: half_adder use entity WORK.half_adder (behavioral);
for H2: half_adder use entity WORK.half_adder (structural);
for O1: or_2 use entity POWER.lpo2 (behavioral)
generic map(gate_delay => gate_delay)
port map (I1 => x, I2 => y, Z=>z);

begin      -- component instantiation statements
H1: half_adder port map (x =>In1, y => In2,
                               sum => s1, carry=> s2);
H2: half_adder port map (x => s1, y => c_in,
                               sum => sum, carry => s2);
O1: or_2 port map(x => s2, y => s3, z => c_out);

end architecture structural;
```

Library Name

Entity Name

Architecture
Name

FIGURE 5.21   An example of using configuration specifications for the structural model of a full adder

Thus, we see that we can specify any type of binding, as long as the ports and arguments match up. We do not even need to use the same names. However, just as wiring up a circuit becomes a bit more involved, so does the writing and management of the models. The readability of the code is particularly important as the sizes of the models grow; the choice of names then becomes important also. If we keep the same names for the entities and the components that are bound to them, then we can rely on default rules for configuring the simulation models, and no **port map** clause is required in the configuration statement.

There is some economy of expression that we can employ if all of the half_adder components in the circuit use the same entity–architecture pair. In this case, we can replace the two component instantiation statements corresponding to H1 and H2 by a single component instantiation statement:

for all: half_adder **use entity** WORK.half_adder (behavioral);

We can think of configuration specifications as syntactic representations of what we might ask for verbally if we were wiring up the circuit on a protoboard. For example, the preceding configuration statement can be thought of as stating "For all of the half-adder components, use the same chip XYZ." Machine-readable representations of such statements must follow a precise syntax and have clear semantics associated with the statements.

5.6.3   Configuration Declaration

The configuration specification is part of the architecture; hence, we must be place it within the architecture body. Modification of our choice of models to implement a component requires editing the architecture and recompiling the model. A configuration declaration enables us to provide the same configuration information, but as a separate design unit and, if desired, in a separate file. In the same way that entities and architectures are design units, so are configuration declarations. Suppose we take all of the configuration information provided in the architecture in Figure 5.21, name it, and refer to it by its name. This unit is a configuration declaration and is a distinct design unit. Figure 5.22 gives an example of the configuration information in Figure 5.21, provided as a configuration declaration.

```
configuration Config_A of full_adder is   -- name the configuration
                                           -- for the entity
for structural   -- name of the architecture being configured
for H1: half_adder use entity WORK.half_adder (behavioral);
end for;

--

for H2: half_adder use entity WORK.half_adder (structural);
end for;

--

for O1: or_2 use entity POWER.lpo2 (behavioral)
generic map(gate_delay => gate_delay)
port map (I1 => a, I2 => b, Z=>c);
end for;

--

end for;
end configuration Config_A;
```

FIGURE 5.22   A configuration declaration for the structural model of the full adder in Figure 5.21.

As with other design units, we name configuration declarations as shown on the first line. In addition, the entity that is to utilize the configuration information is also named. Note that this declaration looks very similar to an architecture declaration. The second line identifies the name of the specific architecture of the entity that is being configured. For example, there could be another structural model of the full adder placed in an architecture labeled Structural_B. We must be able to distinguish between alternative architectures unambiguously and do so by referring to the unique architecture labels.

A close examination of the syntax reveals that the **for** statements are terminated by **end for** clauses. While the preceding declaration deals with only one level of the hierarchy, we can write configuration declarations to span a complete design hierarchy with nested **for**...**end for** constructs to bind components at all levels of the hierarchy. It is also apparent that we can have different configurations for the full adder. For example, we might have a Config_B and a Config_C. Each configuration could use a different set of components or models for the half adder and two-input OR gate components. We might be motivated to take this approach to study the implementation with different technologies—for example, low-power vs. high-speed implementations.

There are several other advanced topics in the area of binding components to architectures, such as direct instantiation, incremental binding, and binding to configurations rather than an entity–architecture pair. These topics can be found in any advanced text on VHDL. The key issue to be understood here is that configurations are the language mechanism that specifies a particular implementation when a myriad of alternative models is available for the constituent components. The use of configurations is motivated in part by the need to be able to reuse models and share models among developers. This implies that it should be easy to selectively replace individual components of large simulation models, and configurations are the VHDL solution to this problem.

Simulation Exercise 5.4: Use of Configurations

This exercise will emphasize the need and importance of configurations. The exercise builds on Simulation Exercise 5.2, which produced two distinct models of an 8-bit adder. The first model was a structural model hierarchically built from smaller size ALUs. The second was a behavioral model constructed with the use of processes.

Step 1.   Construct a 16-bit ALU from two 8-bit ALUs using ripple carry. Use the entity description of the 8-bit ALU developed in Simulation Exercise 5.2.

Step 2.   In the model, include a configuration specification, such as the one shown in Figure 5.21, to specify the name of the architecture of the 8-bit model that you wish to use. Start with the hierarchical 8-bit model. All of the models that are used in building this 8-bit ALU are assumed to have been compiled into your working directory. Make sure that the library WORK is set to your current working directory.

Step 3.   Compile the 16-bit ALU. Test the model and ensure that it is functioning correctly.

Step 4.   Now modify the configuration specification to use the behavioral model of the 8-bit ALU. This should require editing one line (in fact one word) of your VHDL model—the line in your configuration specification.

Step 5.   Test the model and ensure that it is working. Note the ease with which it is possible to "plug in" different models of subcomponents of the 16-bit ALU.

Step 6.   List some of the differences between the two models that you have constructed.

End Simulation Exercise 5.4

5.7   Common Programming Errors

The following are some common programming errors:

- Modifying the model of a component and forgetting to reanalyze the component model prior to reuse.

- Generics can have their values defined at three places: within the model, in a component instantiation statement using the **generic map()** construct, and within an architecture in a component declaration. Changing the value of the generics in one place may not have the intended effect, due to precedence of the other declarations. The actual value of the generic parameter may not be what you expect.

- In using default bindings of components, the name, type, and mode of each signal in the component declaration must exactly match that of the entity; otherwise an error will result.

- Inheriting a generic value by way of default initializations in the component declarations in the higher level may lead to unexpected values of the generic parameters. A clear idea of how generic values are propagated through the hierarchy is necessary to ensure that the acquired values of the generic parameters correspond to the intended values.

- In instantiating components, you may overlook a signal in a component. This does not necessarily lead to compiler errors, and the signal is modeled as unconnected in the design. The result can be incorrectly computed values or undefined signals propagating through the design.

5.8   Chapter Summary

The focus of this chapter has been on the ability to specify hierarchical models of digital systems, *ignoring* how we may specify the internal behavior of components. We can describe internal behavior with language features described in Chapter 3 and Chapter 4. An important aspect of the construction of hierarchical models is the capability to construct parameterized behavioral models and to be able to determine values of parameters by passing information down this hierarchy. This capability is facilitated by the **generic** construct and enables the construction of libraries of models that can be shared by designers. Finally, given such a library of alternative models for a component, the **configuration** construct specifies the particular models to be used in the construction of a model.

The concepts covered in this chapter include the following:

- Structural models
  - component declaration
  - component instantiation
- Construction of hierarchical models
  - abstraction
  - trade-offs between accuracy and simulation speed
- Generics
  - specifying generic values
  - constructing parameterized models
- Generate statement
  - compact descriptions of structured arrays of components
- Configurations
  - component binding
  - default binding rules
  - configuration specification
  - configuration declaration

We now have command of the basic constructs for creating VHDL models of digital systems. The chapters that follow address remaining issues in support of these basic constructs, to provide a complete set of modeling constructs.

Exercises

1. Complete the structural model of the bit-serial adder shown in Figure 5.4 by constructing a model for the two components. You must complete the design of the combinational logic component. Compile, simulate, and test the model.

2. Consider the detailed hierarchical model of an 8-bit ALU constructed in Simulation Exercise 5.2. Now consider a single-level behavioral model of the 8-bit ALU constructed using a single process and 8-bit data types. Compare the two models and comment on the simulation accuracy, simulation time, and functionality.

3. You are part of a software group developing algorithms for processing speech signals for a new digital signal processing chip. To test your software, your options

are (i) to construct a detailed hierarchical model of the chip made up of gate-level models at the lowest level of the hierarchy and (ii) to construct a behavioral-level model of the chip that can implement the algorithms that you wish to use. Your goal is to produce correct code for a number of algorithms prior to detailed testing on a hardware prototype. How would you evaluate these choices, and what are the trade-offs in picking one approach over the other?

4. Construct and test a structural model of the accompanying circuit. Note that there are many different ways in which to do this. You might consider each gate a component or groups of gates as a component represented by the Boolean function computed by those gates.

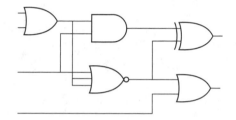

5. Consider the accompanying circuit. Construct a structural model composed of two components: a generic N-input AND gate and a two-input OR gate. By passing the appropriate generic value, we can instantiate the same basic AND gate component as a two-input or three-input AND gate.

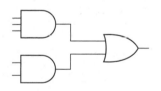

6. In problem 5, use generics to set default gate delays for the components. Now instantiate each AND gate with different gate delays, using the **generic map** construct.

7. Modify the generic model of an N-bit register shown in Figure 5.13 to operate as a counter that is initialized to a preset value.

8. Compare the use of configuration specifications and configuration declarations. When is one or the other advantageous?

9. Implement the structural model of a full adder by using configuration specifications as shown in Figure 5.21. Use a simple model of the two-input OR gate rather than the one shown in the figure. You can omit the configuration statement for the OR gate and use the default binding for this component. Use two different architectures for the half-adder components.

10. Using the half-adder description from Section 5.1 as the basic building block, develop a structural model that implements the following functions: (i) A+B+C and (ii) AB+AC+BC

11. Write and test a VHDL model for a 4×16 decoder assuming that a model for a 3×8 decoder with the entity description given below exists:

entity decoder_3bit **is**
port (input: **in** std_logic_vector(2 **downto** 0);
enable : **in** std_logic;
z : **out** std_logic_vector(7 **downto** 0)
);
end entity decoder_3bit;

12. We have seen how to implement a ripple carry adder in VHDL. Now consider the implementation of an 8-bit carry look-ahead adder.

CHAPTER 6   Subprograms, Packages, and Libraries

With any large body of software, we need mechanisms for structuring programs, reusing software modules, and otherwise managing design complexity. In conventional languages, mechanisms for doing so have been available to us for some time. The VHDL language also provides support for such mechanisms through the definition and use of procedures and functions for encapsulating commonly used operations, as well as the concepts of packages and libraries for sharing large bodies of code.

However, hardware description languages possess several attributes that do not have counterparts in conventional programming languages. For example, the presence of the **signal** class of objects and the notion of simulation time are very different abstractions from those we normally use. The VHDL models may be used for the discrete event simulation of physical systems or to generate an implementation of the physical system described by the model. Collectively, these features generate considerations that do not arise in conventional programming languages. For example, can **wait** statements be used in a procedure? How are signals passed as parameters to procedures and modified? Can functions operate on signals? The essential issues governing the use of functions and procedures are discussed initially in this chapter.

Related groups of functions and procedures can be aggregated into a module that can be shared across many different VHDL models. Such a module is referred to as a *package*. In addition to the definitions of procedures and functions, packages may contain user-defined data types and constants. Packages in turn can be placed in *libraries*. Libraries are repositories for design units in general, and packages are one type of design unit. Other design units are entities, architectures, and configurations. Collectively, procedures, functions, packages, and libraries provide facilities for creating and maintaining modular and reusable VHDL programs.

6.1   Essentials of Functions

As in traditional programming languages, functions compute a value based on the values of the input parameters. An example of a function declaration is

> **function** rising_edge (**signal** clock: **in** std_logic) **return boolean**;

The function definition provides a function name and specifies the input parameters and the type of the result. Functions return values that are computed using the input parameters. Therefore, we would expect that the parameter values are used, but not changed within the function. This notion is captured in the **mode** of the parameter. Parameters of mode **in** can only be read. Functions cannot modify parameter values (procedures can), and therefore functions do not have any parameters of mode **out**. Since the mode of all function parameters is **in**, we do not have to specify the mode of a parameter. Section 6.2 discusses other possible modes with a discussion of procedures.

Consider the structure of a function as shown in Figure 6.1. The function has a name (rising_edge) and a set of parameters. The parameters in the function definition are referred to as *formal* parameters. Formal parameters can be thought of as placeholders which describe the type of object that will be passed into the function. When the function is actually called in a VHDL module, the arguments in the call are referred to as *actual* parameters. For example, the preceding function may be called in the following manner:

> rising_edge (enable);

In this case, the actual parameter is the signal enable, which takes the place of the formal parameter clock in the body of the function. The type of the formal and actual parameters must match—except for formal parameters that are constants. In this case, the actual parameter may be a variable, signal, constant, or expression. When no class is specified, the default class of the parameter is constant. Wait statements are not permitted in functions. Thus, functions execute in zero simulation time. It follows that wait statements cannot exist in any procedures called by a function (although procedures are allowed to have wait statements). Furthermore, parameters are

```
function rising_edge (signal clock: std_logic) return boolean is
--
--declarative region: declare variables local to the function
--
begin
--
-- body
--
return (value)
end function rising_edge;
```

FIGURE 6.1   Structure of a function

restricted to be of mode **in**, and therefore functions cannot modify the input parameters. Thus, signals passed into functions cannot be assigned values. This behavior is
consistent with the conventional definition of functions.

'87 vs. '93

☞

VHDL'93 supports two distinct types of functions: pure functions and impure
functions. The former are functions that always return the same value when called with
the same parameter values. Such functions conform to what we normally regard as the
mathematical definition of a function. Impure functions, on the other hand, can return
different values when called with the same parameter values at different times. This is
possible because functions may have visibility over signals that are not in the parameter list—for example, ports of the encompassing entity. Pure functions occur commonly as type conversion and resolution functions. In this text we restrict our discussion
to pure functions.

Example: Detection of Signal Events

Often, we find it useful to perform simple tests on signals to determine whether certain events have taken place. For example, the detection of a rising edge is common in
the modeling of sequential circuits. Figure 6.2 shows the VHDL model of a positive

```
library IEEE;
use IEEE.std_logic_1164.all;
entity dff is
port (D, Clk : in std_logic;
      Q, Qbar : out std_logic);
end entity dff;

architecture behavioral of dff is
function rising_edge (signal clock : std_logic) return boolean is
variable edge : boolean:= FALSE;
begin
edge := (clock = '1' and clock'event);
return (edge);
end function rising_edge;

begin
output: process is
begin
wait until (rising_edge(Clk));

    Q <= D after 5 ns;
    Qbar <= not D after 5 ns;

end process output;
end architecture behavioral;
```

FIGURE 6.2   An example of the use of functions

edge-triggered D flip-flop from Figure 4.10. The only difference is the inclusion of a function for testing for the rising edge, rather than having the function code in the body of the VHDL description. Note the placement of the function in the declarative portion of the architecture. Normally, we use this region to declare signals and constants that appear in the body of the architecture. Therefore, we might expect that we can also declare functions (or procedures) that are used in the architecture as well. This is indeed the case. The function could also have been declared in the declarative region of the process that called the function (i.e., between the keywords **process** and **begin**.) The question is whether you wish to have the function be visible to, and therefore callable from, all processes in the architecture body, or whether it should be visible to just one process. In practice, we would much rather place related functions and procedures in packages, a type of design unit described later in the chapter.

Example End: Detection of Signal Events

6.1.1 Type Conversion Functions

Type conversion is another common instance of use functions. The model of the memory module in Section 4.1 represented memory as a one-dimensional array indexed by an integer. However, memory addresses are provided as n-bit binary addresses. We find that we need to convert this bit vector representing the memory address to an integer used to index the array representing memory. In other instances, we may want to use models of components developed by others. We would most likely use their models by instantiating their components as part of a larger structural model. For example, suppose we are designing an arithmetic–logic unit and wish to incorporate a multiplier developed by a colleague. We do not need to understand the internal operation of the multiplier, but we do need to know the input and output signals and their types so that we may correctly interface to the multiplier. Let us say that our colleague has used signals of type **bit** and **bit_vector**, while we have been using signals of type std_logic and std_logic_vector. Then type conversion functions will be necessary for interoperability if we do not wish to invest in the time required to convert their models to use the IEEE 1164 types.

Example: Type Conversion

Consider the VHDL type **bit_vector** and the IEEE 1164 type std_logic_vector. We may wish to make assignments from a variable of one type to a variable of the other type. For example, consider the conversion of a signal of type **bit_vector** to std_logic_vector. We can use the function to_stdlogicvector() that is provided in the package std_logic_1164.vhd (more on packages in Section 6.4). This function takes as an argument an object of type **bit_vector** and returns a value of type std_logic_vector. Conversely, we may wish to convert from std_logic_vector to **bit_vector**. Figure 6.3

```
function to_bitvector (svalue : std_logic_vector) return bit_vector is
variable outvalue : bit_vector (svalue'length-1 downto 0);
begin
for i in svalue'range loop -- scan all elements of the array
case svalue (i) is
when '0' => outvalue (i) := '0';
when '1' => outvalue (i) := '1';
when others => outvalue (i) := '0';
end case;
end loop;
return outvalue;
end function to_bitvector;
```

FIGURE 6.3   An example of a type conversion function

gives an example of the implementation of this function. Recall that each element in a std_logic_vector is of type std_logic and each element in a **bit_vector** is of type **bit**. The function simply scans the vector and converts each element of the input std_logic_vector to a **bit_vector** element. The type **bit** may take on values 0 and 1, whereas the type std_logic may take on one of nine values. Note the declaration of the variable outvalue. The function declaration does not provide the size of the number of bits in the argument. This is set when the formal parameter is associated with the actual parameter at the time the function is called. In this case, how can we declare the size of any local variable that is to have the same number of bits as the input parameter? The answer is, by using attributes. As discussed in Section 4.5, arrays have an attribute named **length**. The value of svalue'**length** is the length of the array. By using unconstrained arrays in the definition of the function, we can realize a flexible function implementation wherein the actual sizes of the parameters are determined when the actual parameters are bound to formal parameters, which occurs when the function is called.

There are many ways in which to perform such type conversions, and Figure 6.3 shows but one of them. By examining commercial packages such as std_logic_arith.vhd and std_logic_1164.vhd, we will find many such conversion functions. Examples include conversion from std_logic_vector to **integer** and vice versa. Check the libraries that come with the installation of your VHDL simulator and tools. You will find many packages, and it is useful to browse through them and study the procedures and functions that are contained within them.

Example End: Type Conversion

6.1.2   Resolution Functions

Resolution functions comprise a special class of functions. Recall from Chapter 4 that the resolved type is a signal that may have multiple drivers. This occurs quite often in

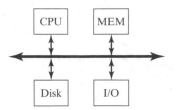

FIGURE 6.4   A simple model of a computer

digital systems. For example, consider a high-level model of a computer system, shown in Figure 6.4. The CPU, memory, and some peripherals, such as a disk and other I/O devices, must exchange data. It is quite expensive to have a dedicated interconnect between every pair of communicating devices. Furthermore, it is not necessarily the case that all of the devices have to communicate at the same time. A common architecture is to have these devices communicate over a shared set of signals called a bus. In the VHDL model of such an architecture, the shared bus could be a signal datatype, and several components may make assignments to the shared signal during the course of a simulation. The physical analogy of such assignments is that of multiple physical drivers placing values on this bus at different points in time.

Circuits that implement wired logic are another example of instances in which a signal can have multiple drivers. The design is such that the final signal value represents a logical operation such as a Boolean AND. The next example illustrates the need for shared signals.

Example: Programmable Array Logic

Switches are often used to implement logic gates as shown in Figure 6.5. The switches are turned on by a 0 value on the control input. A pull-up circuit normally drives the output signal to a logic high, or 1, value. When a switch is turned on, it pulls the output signal down to a logic low, or 0, value. Thus, in order for the output to be at a logic 1 value, all of the switches must be off. If any one of the switches is turned on, the output signal is pulled to a logic 0. This behavior is the logic AND function. Now imagine describing the behavior of this circuit in VHDL. We can conceive of an architecture description wherein the output signal is declared in the architecture. The operation of each switch might be described by a process that is sensitive to the value of its control input signal and drives the output signal accordingly. Thus, each process will drive the output signal. The actual value of the signal is determined by checking to see whether any of the processes are driving the signal to a logic 0. The function that checks all of the values being driven by each process is the resolution function. In this case, the resolution function returns a value of 0 for the signal if any of the processes attempts to drive the signal to a value of 0.

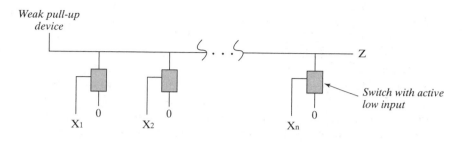

Any $X_n = 0$ turns switch on, producing $Z = 0$.

FIGURE 6.5   An example of wired AND logic

In general, the VHDL model should be capable of handling more general cases. For example, we should be able to describe other ways to pick the value of the shared signal when multiple drivers try to schedule values on the shared signal at the same point in simulation time.

Example End: Programmable Array Logic

To correctly simulate circuits such as those described in the preceding example, we must be able to unambiguously state the value of the signal at any point in time. The value must be *resolved* on the basis of the values scheduled by the multiple drivers of the signal. The algorithm for resolving the issue of the signal value at any time is captured in the *resolution function*. One can think of the resolution function as examining all of the scheduled values on the bus for that time and determining the value of the signal. For example, the value could be a logical OR of all of the signals or the maximum value.

The shared signal must be declared as a *resolved type*, which means that there is a resolution function associated with all signals of this type. This function should accurately reflect the behavior of the physical system being modeled. During the course of a simulation, when any signal of this type is to be assigned a value, the resolution function is invoked. The function examines the values on all drivers for that signal and computes the correct signal value, as defined by the resolution function. The resolution function must be an associative operation so that the order in which the multiple signal drivers' values are examined does not affect the resolved value of the signal. For example, a logical OR operation on a set of values of type **bit** is an associative operation. Similarly, the logical AND function, as well as maximum and minimum value functions, are associative operations. Throughout this text, we have used the signal type std_logic—a resolved type defined by the IEEE 1164 standard. The next example, from the implementation of the IEEE 1164 standard, illustrates how resolved types can be declared and how resolution functions can be defined.

Example: Resolved Types in the IEEE 1164 Standard

Let us examine the definition and use of the resolved type std_logic from an implementation of the IEEE 1164 standard. Figure 6.6 shows an example of the declaration and use of resolved types taken from an implementation of the IEEE 1164 standard, std_logic_1164.vhd, that is provided with just about any VHDL toolset. This implementation of the standard first defines a new type: std_ulogic. A signal of this type takes on nine values, as defined in Figure 6.6 and as described in Section 2.5. The type is referred to as an enumerated type, since the values of an object of that type are explicitly enumerated. However, any signal declared to be of that type can support only a single driver. Therefore, we wish to define a new signal type that can take on all of the values of std_ulogic, but can also support multiple drivers. This is done on the following line by creating the (sub) type std_logic.

Consider the structure of this declaration. It looks very much like any other declaration, except for two items. A *subtype* simply means that the declared signal can take on a range of values that is a subrange of the original or *base* type. Second, the type provides the name of a *resolution function* that is associated with all objects declared to be of that type. In this case, it is the function named resolved. What this definition means is the following: Whenever a signal of type std_logic is to be assigned a value, there may be multiple drivers associated with the signal. The values from these multiple drivers are passed to the resolution function, which determines the value to be assigned to the signal. For example, consider the case in which one driver drives a single-bit bus to a logic 1 while another driver leaves the bus in a high-impedance state or Z. The value of the signal should be a logic 1. The resolution function must be capable of making this determination and of handling more than two drivers.

One simple approach to implementing a resolution function is to build a table. The row and column indices correspond to the signal values from two drivers. The table entry corresponds to the value that would be produced if the two drivers were

```
type std_ulogic is ('U', -- Uninitialized
                    'X', -- Forcing Unknown
                    '0', -- Forcing 0
                    '1', -- Forcing 1
                    'Z', -- High Impedance
                    'W', -- Weak Unknown
                    'L', -- Weak 0
                    'H', -- Weak 1
                    '-' -- Don't care
                );

function resolved (s : std_ulogic_vector) return std_ulogic;

subtype std_logic is resolved std_ulogic;
```

FIGURE 6.6 An example of the declaration of resolved signals

TABLE 6.1   Table for resolving the values of a pair of signals of type std_logic

	U	X	0	1	Z	W	L	H	-
U	U	U	U	U	U	U	U	U	U
X	U	X	X	X	X	X	X	X	X
0	U	X	0	X	0	0	0	0	X
1	U	X	X	1	1	1	1	1	X
Z	U	X	0	1	Z	W	L	H	X
W	U	X	0	1	W	W	W	W	X
L	U	X	0	1	L	W	L	W	X
H	U	X	0	1	H	W	W	H	X
-	U	X	X	X	X	X	X	X	X

attempting to drive a signal to those two values. For example, the entry in a table at location (Z,1) would be 1. Now, if we had a set of drivers, we could compute the final value by resolving the values of all drivers in a pairwise manner—hence the requirement for associativity. Table 6.1 shows the structure of the table used by an implementation of the IEEE 1164 standard to resolve the values of a pair of signals of type std_logic. For example, two driver values of Z and W will yield a signal value of W.

Example End: Resolved Types in the IEEE 1164 Standard

Example: Using Resolution Functions

Let us consider another example of the implementation of resolution functions. Consider a system with multichip modules (MCMs): i.e., chip carriers that can support multiple semiconductor dies on a single substrate and within a single package. Suppose that we have multiple dies placed on a single multichip module, as shown in Figure 6.7. Assume that, for this MCM module to be functional, all of the dies must be functional. This means that, when one die has failed, the module is considered defective. Periodically, self-test

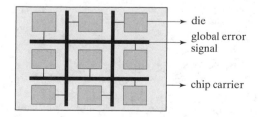

FIGURE 6.7   Structure of a multichip module with multiple dies

circuitry in the die (or even in the MCM substrate!) will locally run diagnostics to test the individual dies to ensure that they remain functional. There is a single global error signal that is driven by an output of the self-test circuitry within each die. This global signal performs the logical OR of all of the single-bit diagnostic results from the individual die. If this global error signal is asserted, then at least one bad die exists on the MCM and the complete package is considered to be faulty. Since there are multiple drivers, this global signal must be of a resolved type. Figure 6.8 shows an example of the VHDL code that may be used to implement this signal type.

In this model, we assume that the behavior of each chip is captured with a process. Each process may place a value of 0 or 1 on the shared signal error_bus. If at least one driver forces error_bus to 1, we would like the value of error_bus to remain 1. This behavior is captured in the resolution function shown in Figure 6.8.

```
library IEEE;
use IEEE.std_logic_1164.all;

entity mcm is -- an empty entity description
end entity mcm;

architecture behavioral of mcm is
function wire_or (sbus :std_ulogic_vector) return std_ulogic;
begin
for i in sbus'range loop -- this loop implements a logical OR across all signals
 if sbus(i) = '1' then
return '1';
end if;
end loop;
return '0';
end function wire_or;

subtype wire_or_logic is wire_or std_ulogic; -- declare the new resolved type
signal error_bus : wire_or_logic; -- this signal is global to all processes
begin
Chip1: process is                    There could be many more
begin                                processes like this, e.g.,
-- ..                                corresponding to each die.
error_bus <= '1' after 2 ns;
-- ..
end process Chip1;
Chip2: process is
begin
-- ..
error_bus <= '0' after 2 ns;
-- ..
end process Chip2;
end architecture behavioral;
```

FIGURE 6.8   An example of the use of resolution functions

The function wire_or receives an array of values as input. The loop simply scans the array, looking for the first 1. One very interesting feature about this function is that the parameter is an unconstrained array. This means that we have not specified the number of elements that will be passed to the function. Can we pass an array of 32 elements? 64 elements? The answer is yes to both. The size of the input parameter will be determined at the time the function is called. In this case, we do not know how many signal drivers may exist. The use of unconstrained arrays in this manner is quite common. This body of the function simplifies the process of resolving signal values. The base type of wire_or_logic is std_ulogic. Thus, any signal of type wire_or_logic can actually take on values other than 1 or 0. However, the resolution function shown here reflects a classical 0/1 view of single-bit signals and ignores these other values. A more realistic and robust approach would account for all possible combination of signal values.

Example End: Using Resolution Functions

The second class of subprograms consists of procedures. A distinguishing feature of procedures is that they can modify input parameters. In this case, we must consider how signals are passed and handled. The next section addresses the essential issues in getting started with writing and using procedures.

6.2   Essentials of Procedures

Procedures are subprograms that can modify one or more of the input parameters. The following procedure declaration illustrates the procedure call interface:

>   **procedure** read_v1d (**variable** fname: **in text**; v : **out** std_logic_vector);

This is a procedure to read data from a file, where fname is a file parameter. The first characteristic we might notice is that parameters may be of mode **out**. Just as parameters of mode **in** must be read and cannot be written, parameters of mode **out** cannot be read and used in a procedure, but can only be written by the procedure. We may also have parameters of mode **inout**. As with functions, the type of the formal parameters in a procedure declaration must match the type of the actual parameters used when the procedure is called. If the class of the procedure parameter is not explicitly declared, then parameters of mode **in** are assumed to be of class **constant**, while the parameters of mode **out** or **inout** are assumed to be of class **variable**. Each call to a procedure initializes variables declared within a procedure, and their values do not persist across invocations of the procedure.

Example: Interface to Memory

Let us consider a VHDL model for a simple processor in which we have two components: a CPU and memory. The behavioral model of the CPU must be able to read and write locations from memory. These operations are common candidates for implementation as procedures. We will create two procedures: one to read, and one to write, memory locations. We will assume the memory model and associated signals as shown in Figure 4.1, with the addition of one additional signal from memory that signifies the completion of a memory operation. Figure 6.9 shows these procedures; both should

```
library IEEE;
use IEEE.std_logic_1164.all;
entity CPU is
port (write_data : out std_logic_vector (31 downto 0); -- data from memory
    ADDR :out std_logic_vector (2 downto 0); -- CPU generated address
    MemRead, MemWrite: out std_logic; -- read and write control signals from CPU
    read_data : in std_logic_vector (31 downto 0); -- data to memory
    S : in std_logic);
end entity CPU;

architecture behavioral of CPU is
procedure mread (address : in std_logic_vector (2 downto 0);
                signal R : out std_logic;
                signal S : in std_logic;
                signal ADDR: out std_logic_vector (2 downto 0);
                signal data : out std_logic_vector (31 downto 0)) is
begin
ADDR <= address;
R<= '1';
wait until S = '1';
data <= read_data;
R<= '0';
end procedure mread;

procedure mwrite (address : in std_logic_vector (2 downto 0);
                signal data : in std_logic_vector (31 downto 0);
                signal ADDR : out std_logic_vector (2 downto 0);
                signal W : out std_logic;
                signal DO : out std_logic_vector (31 downto 0)) is
begin
ADDR <= address;
DO<= data;
W<= '1';
wait until S = '1';
W <= '0';
end procedure mwrite;
--
-- any signal declarations for the architecture here
--
```

FIGURE 6.9   An example of the use of procedures

```
begin
--
-- CPU behavioral description here
--
process is
begin
--
-- behavioral description
--
end process;

process is
begin
--
-- behavioral description
--
end process;
end architecture behavioral;
```

FIGURE 6.9   (Continued)

manipulate the signals illustrated in the memory interface in Figure 4.1. There are a number of interesting features here. Note the presence of **wait** statements within the procedure. Thus, a process can suspend execution inside a procedure. Furthermore, signals can be assigned values within a procedure—a capability that raises the issue of how signals are passed into the procedure, and this issue is dealt with in Section 6.2.1. Signals that are modified within the procedure are declared to be of mode **out**. For example, see signal R in procedure mread() in Figure 6.9.

The body of the architecture description is likely to include processes within which procedures can be called. Alternatively, the procedures could have been declared within the declarative region of the process, just before the **begin** statement and after the **process** statement. Just as processes can declare and use variables that are local to the process, processes may also declare and use procedures within a process. However, in this case, they would be visible only within that process.

Example End: Interface to Memory

6.2.1   Using Procedures

Signals cannot be declared within procedures. However, signals can be passed into procedures as parameters. Due to visibility rules, procedures can make assignments to signals that are not explicitly declared in the parameter list. For example, procedures declared within a process can make assignments to signals corresponding to the ports of the encompassing entity. This is possible because the ports are visible to the process. The procedure is said to have side effects, since it has an effect on a signal that is not declared in the parameter list. This is poor programming practice, since it makes it difficult to reason about the models (e.g., when debugging) and understand their behavior. Clarity and understanding of the code are enhanced if parameters are passed explicitly, rather than relying on side effects. If the class of a parameter is not declared, and the mode is **out** or **inout**, then the class defaults to that of a **variable**. If the mode is **in**, the class of the parameter defaults to a **constant**.

Procedures can also be placed in the declarative region of a process. We know that processes cannot have a sensitivity list and also have wait statements in their bodies. Therefore, when we use procedures, it follows that a process that calls a procedure with a **wait** statement cannot have a sensitivity list.

6.2.2   Concurrent and Sequential Procedure Calls

Depending on how procedures are used, we can distinguish between concurrent and sequential procedure calls. Recall the concurrent signal assignment statements from Section 3.3.1. Each statement represented the assignment of a value to a signal, and this assignment occurred in simulated time concurrently with the execution of the other concurrent signal assignment statements and processes. Concurrent procedure calls can be viewed similarly. The procedure is invoked in the body of an architecture concurrently with other concurrent procedures, concurrent signal assignment statements, or processes. The procedure is invoked when there is an event on a signal that is an input parameter to the procedure. It follows that if we use a concurrent procedure, the parameter list cannot include a variable (in VHDL'87), since variables cannot exist outside of a process. However, VHDL'93 supports shared variables, but we consider them to be advanced features not discussed here. In contrast, sequential procedure calls are those which invoke the procedure within the body of a process. In this case, the sequence of execution of statements within the process determines the invocation of the procedure just as in a conventional program. The next example should help solidify our understanding of the differences between concurrent and sequential procedure calls.

'87 vs. '93
☞

Example: Concurrent and Sequential Procedure Calls

Figure 6.10 illustrates an example of a concurrent procedure call. The structural model of a bit-serial adder from Figure 5.4 has been rewritten such that a procedure replaces the D flip-flop component instantiation. The procedure is invoked concurrently with the component comb whenever there are events on the signals that are declared to be of mode **in**. Thus, events on the clk, reset, or d inputs will cause this procedure to be invoked. From the procedure body, we see that the output is modified only on the rising edge of the clk signal.

```
library IEEE;
use IEEE.std_logic_1164.all;
entity serial_adder is
port (a, b, clk, reset : in std_logic;
        z : out std_logic);
end entity serial_adder;

architecture structural of serial_adder is
component comb is
 port (a, b, c_in : in std_logic;
        z, carry : out std_logic);
end component comb;

procedure dff(signal d, clk, reset : in std_logic;
                    signal q, qbar : out std_logic) is
begin
 if (reset = '0') then
    q <= '0' after 5 ns;
    qbar <= '1' after 5 ns;
    elsif (rising_edge(clk)) then
    q <= d after 5 ns;
    qbar <= (not d) after 5 ns;
 end if;
end procedure dff;

signal s1, s2 : std_logic;

begin
C1: comb port map (a => a, b => b, c_in => s1, z =>z, carry => s2);
--
-- concurrent procedure call
--
dff(clk => clk, reset =>reset, d=> s2, q=>s1, qbar =>open);
end architecture structural;
```

FIGURE 6.10   An example of a concurrent procedure call

This structure does appeal to our understanding of VHDL programs. Consider what would happen in the model in the example of Figure 5.4. Let us assume that when this model is simulated, the component dff will be replaced by a behavioral model similar to the one shown in Figure 4.10. We see that the procedure effectively implements the same behavior. Note how the parameter list explicitly associates the formal and actual parameters, rather than having this association made by virtue of the position in the call. The signal qbar is associated with the keyword **open** in the procedure call. This is akin to leaving a pin of a device (dff) unconnected.

```
library IEEE;
use IEEE.std_logic_1164.all;
entity serial_adder is
port (a, b, clk, reset : in std_logic;
      z : out std_logic);
end entity serial_adder;

architecture structural of serial_adder is
component comb is
 port (a, b, c_in : in std_logic;
       z, carry : out std_logic);
end component comb;
procedure dff(signal d, clk, reset : in std_logic;
                    signal q, qbar : out std_logic) is
begin
 if (reset = '0') then
     q <= '0' after 5 ns;
     qbar <= '1' after 5 ns;
   elsif (clk'event and clk = '1') then
     q <= d after 5 ns;
     qbar <= (not d) after 5 ns;
  end if;
end procedure dff;
signal s1, s2 : std_logic;
begin
C1: comb port map (a => a, b => b, c_in => s1, z =>z, carry => s2);
process
begin
 dff(clk => clk, reset =>reset, d=> s2, q=> s1, qbar => open);
wait on clk, reset, s2;
end process;
end architectural structural;
```

FIGURE 6.11   An example of a sequential procedure call

Figure 6.11 shows the equivalent implementation as a sequential procedure call. The procedure is encased in a process with an explicit **wait** statement. Note the structure of the **wait** statement. If an event occurs on any of the signals in the list, the process will be executed, which in this case will cause the procedure dff() to be called. This procedure call model is equivalent to the model shown in Figure 6.10.

Example End: Concurrent and Sequential Procedure Calls

6.3   Subprogram and Operator Overloading

A very useful feature of the VHDL language is the ability to *overload* the subprogram name. For example, there are several models and implementations of a D flip-flop. We saw a few examples in Section 4.4. Imagine that we were to write behavioral models of sequential circuits that included D flip-flops. We might be using procedures such as the one shown in Figure 6.10 to model the behavior of a D flip-flop. If we wished to incorporate models that had asynchronous set and clear signals, we might write another procedure with a different name, say asynch_dff(). What if we wish to have procedures that would operate on signal arguments of type **bit_vector** rather than std_logic_vector? Then we would write distinct procedures to incorporate models with these types and behaviors. By accommodating various possibilities of argument types and flip-flop behavior, we might have to write many different procedures, while keeping track of the names to distinguish them.

It would be very helpful if we could use a single name for all procedures describing the behavior of various types of D flip-flops. We would like to call dff() with the right parameters and let the compiler determine which procedure to use based on the number and type of arguments. For example, consider the following two procedure calls:

```
dff(clk, d, q, qbar)
dff(clk, d, q, qbar, reset, clear)
```

From the arguments, we can see that we are referring to two different procedures, one that utilizes asynchronous reset and clear inputs and one that does not (e.g., corresponding to Figure 4.10 and Figure 4.11, respectively). From the type and number of arguments, we can tell which procedure we meant to use. This process is referred to as overloading subprogram names or simply *subprogram overloading*. When we create such a set of procedures or functions with overloaded names, we would probably place them in a package (see Section 6.4) and make the package contents visible via the **use** clause. If we examine the contents of some of the packages shown in Appendix B, we will see examples of overloaded functions and subprograms. For example, note that, in std_logic_1164.vhd, the Boolean functions **and**, **or**, etc., have been defined for the type std_logic.

Similarly, operators such as "*" and "+" have been defined for certain predefined types of the language (e.g., integers). What if we wish to perform such operations on other data types that we may create? We can overload these operators by providing definitions of "*" and "+" for these new data types. CAD tool vendors typically distribute packages that contain definitions of operators and subprograms for various operations on data types that the language does not predefine. For example, the std_logic_arith.vhd package distributed by CAD tool vendors provides definitions for various operators over the std_logic and std_logic_vector types. Two examples of overloading the definitions of "*" and "+" operators taken from this package are the following:

function "*" (arg1, arg2: std_logic_vector) **return** std_logic_vector;
function "+" (arg1, arg2 :signed) **return** signed;

This means that if the contents of this package are included in a model via the **use** clause, then statements such as

s1 <= s1 + s2;

are valid, where all three signals are of type std_logic. Otherwise, the "+" operation is not defined for objects of type std_logic, and the statement would be in error.

Procedures and functions are necessary constructs for building reusable blocks of VHDL code, for hiding design complexity, and for managing large complex designs. As we have seen in this section, they are also a means of enriching the language to easily handle new data types by encapsulating the definitions of common operations and operators over these data types. Even with a small number of new data types, the need to overload all of the common operators can generate quite a large number of functions. Furthermore, when we think of overloading subprogram names, we can generate quite a few additional procedures or functions. Packages are a mechanism for structuring, organizing, and using such user-defined types and subprograms. These concepts are discussed next.

6.4   Essentials of Packages

As we acquire larger groups of functions and procedures within the models that we construct, we must consider how they will be used. We can use text editors and manually insert those functions into the VHDL models as we use them. However, this is a rather tedious process at best, especially when the models we construct grow large. A better approach would be to group logically related sets of functions and procedures into a module that distinct designs and people can easily share. *Packages* are a means for doing so within the VHDL language.

Packages provide for the organization of type definitions, functions, and procedures so that distinct VHDL programs can share them. To gain an intuition for the constructs used in building packages, it is instructive to consider how we try to reuse code modules across projects. When we are working on large class projects, we attempt to make the most efficient use of our time by reusing functions or procedures that we may have written for older programs, found available for free somewhere on the Internet, or garnered from friends. For example, imagine that you have painstakingly put together a package that contains useful functions, procedures, and data types to help designers build simulation models of common computer architectures. This package may include definitions of new types for registers, instructions, and memories, as well as procedures for reading or writing memories, procedures for performing logical shift operations, and functions for type conversion operations. After months of tedious development, you are now interested in promoting the use of the package among fellow VHDL developers. How might you communicate the contents of the package to convince them of its utility? What would developers want or need to know to determine whether they could benefit from using the contents of your package? At the very least, we would need to have a list of the functions and procedures and what they do. For example, for each procedure, what values are computed and returned, and what parameters must be passed to perform these computations. This information forms the basis of the *package declaration*, which in turn forms the interface or specification of the services that your package provides. When we write C or VHDL programs, we must declare the variables or signals that we are using, their type, and possibly initial values. Similarly, when we write packages, we must declare their contents. It is just that their contents are now more complex objects, such as functions, procedures, and data types. The package declaration is the means by which users declare what is available for use by VHDL programs. In the same sense that a hardware design unit possesses an external interface to communicate with other components, the package declaration defines the interface to other VHDL design units.

The easiest way to understand packages is by example, so let us examine a package that provides a new data type and a set of functions that operate on that data type. Throughout this text, the examples have declared and used the package std_logic_1164.vhd. Now let us look inside this package to see how the type is declared and how functions and declarations are used. Figure 6.12 shows a portion of

```
package std_logic_1164 is
    ----------------------------------------------------------------
    -- logic state system (unresolved)
    ----------------------------------------------------------------
    type std_ulogic is ('U', -- Uninitialized
                        'X', -- Forcing Unknown
                        '0', -- Forcing 0
                        '1', -- Forcing 1
                        'Z', -- High Impedance
                        'W', -- Weak Unknown
                        'L', -- Weak 0
                        'H', -- Weak 1
                        '-'   -- Don't care
                    );
    type std_ulogic_vector is array (natural range <>) of std_ulogic;

    function resolved (s : std_ulogic_vector) return std_ulogic;
    subtype std_logic is resolved std_ulogic;

    type std_logic_vector is array (natural range <>) of std_logic;

    function "and" (l, r : std_logic_vector) return std_logic_vector;
    function "and" (l, r : std_ulogic_vector) return std_ulogic_vector;
    --
    --..<rest of the package definition, for example other function and
    -- ..procedure interfaces>
    --
    end package std_logic_1164;
```

FIGURE 6.12 Examples from the package declaration of an implementation of the IEEE 1164 standard

the package declaration of an implementation of the IEEE 1164 package distributed with the vast majority of the VHDL environments. Appendix B.3 gives a listing of the package declaration. We know that the basic VHDL type **bit** can take on only the values 0 and 1 and therefore is inadequate to represent most real systems. By using the concept of enumerated types (see Chapter 9), we define a new type, std_ulogic, as shown in Figure 6.12. Now a signal can be declared to be of this type:

> **signal** *example_signal* : std_ulogic: = 'U';

The signal *example_signal* can now be assigned any one of the nine values previously defined, rather than the two values 0 and 1. The package also declares a resolved type, std_logic, which is a subtype of std_ulogic. This declaration simply states that, in the course of the simulation, when a signal of type std_logic is assigned a value, the resolution function resolved will be invoked to determine the correct value of a signal from the multiple drivers associated with the signal. However, we do have a problem in that all of the predefined logical functions, such as AND, OR, and XOR, operate on signals of type **bit,** which the language predefines. These logic functions must be redefined for signals of the preceding type. Figure 6.12 shows some of these functions, and Appendix B.3 gives their declarations. If we are constructing a package that uses types, procedures, or functions from another package, then we must provide access to this package from our VHDL program via **library** and **use** clauses.

Now that we have defined what is in the package, we must provide the VHDL code that implements these functions and procedures. This implementation is contained in the *package body*, which is essentially a listing of the implementations of the procedures and functions declared in the package declaration. The body is structured as follows:

> **package body** my_package **is**
> --
> -- *type definitions, functions, and procedures*
> --
> **end package body** my_package;

Once we have these packages, how do we use them? Typically they are compiled and placed in *libraries* and referenced within VHDL design units via the **use** clause. All of the examples in this text have utilized the package std_logic_1164.vhd, which is in the library named IEEE. The essential properties of libraries are discussed next.

6.5   Essentials of Libraries

Each design unit—entity, architecture, configuration, package declaration, and package body—is analyzed (compiled) and placed in a *design library*. Libraries are generally implemented as directories and are referenced by a logical name. In the implementation of the VHDL environment, this logical name maps to a physical path to the corresponding directory, and the mapping is maintained by the host implementation. However, just as with variables and signals, before we can use a design library we must declare the library we are using by specifying its logical name. We do this in the VHDL program, using the library clause, which has the following syntax:

library *logical-library-name-1, logical-library-name-2,...;*

In VHDL, the libraries STD and WORK are implicitly declared. Therefore, user programs do not need to declare them. The former contains standard packages provided with VHDL distributions. The latter refers to the working directory, which can be set within the VHDL environment you are using. (Refer to your CAD tool documentation on how this can be done.) However, if a program were to access functions in a design unit stored in a library with the logical name IEEE, then this library must be declared at the start of the program. Most, if not all, vendors provide an implementation of the library IEEE with packages such as std_logic_1164.vhd, as well as other mathematics and miscellaneous packages.

Once a library has been declared, all of the functions, procedures, and type declarations of a package in that library can be made accessible to a VHDL model through the **use** clause. For example, the following statements appear prior to the entity declaration in all of the examples in this text:

library IEEE;
use IEEE.std_logic_1164.**all**;

When these declarations appear just before the **entity** design unit they are referred to as the *context clause*. The second statement in the preceding context clause makes *all* of the type definitions, functions, and procedures defined in the package std_logic_1164.vhd visible to the VHDL model. It is as if all of the declarations had been physically placed within the declarative part of a process that uses them. A second form of the **use** clause applies when only a specific item, such as a function called my_func, in the package is to be made visible.

use IEEE.std_logic_1164.my_func;

The **use** clause can appear in the declarative part of any design unit. Collectively, the **library** and **use** clauses establish the set of design units that are visible to the VHDL analyzer as it is trying to analyze and compile a specific VHDL design unit.

When we first start writing VHDL programs, we tend to think of single entity–architecture pairs when constructing models. We probably organize our files in the same fashion, with one entity description and the associated architecture description in the same file. When this file is analyzed, the **library** and **use** clauses determine which libraries and which packages within those libraries are candidates for finding functions, procedures, and user-defined types that are referenced within the model being compiled. However, these clauses apply only to the immediate entity–architecture pair! *Visibility must be established for other design units separately!*

There are three primary design units: entities, package declarations, and configuration declarations. The context clause applies to the next primary design unit. If we start having multiple design units within the same physical file, then each primary design unit must be preceded by the **library** and **use** clauses necessary to establish the visibility to the required packages. For example, let us assume that the VHDL models shown in Figure 6.2 and Figure 6.8 are physically in the same file. Then the statements

> **library** IEEE;
> **use** IEEE.std_logic_1164.**all**;

must appear at the beginning of *each* model—that is, prior to the **entity** descriptions. We cannot assume that, since we have these statements at the top of the file, they are valid for all design units in the same file. In that case, if we neglected to precede each model with the preceding statements, the VHDL analyzer would return with an error on the use of the type std_logic in the subsequent models, since this is not a predefined type within the language, but rather is defined in the package std_logic_1164, which was not declared for subsequent models.

Simulation Exercise 6.1: Packages and Libraries

This exercise deals with creating and using a simple package for the development of a VHDL simulation model.

Step 1.   Using a text editor, create a package with the following characteristics:

   Step 1(a)   Include several procedures for simulating a D flip-flop. You can start with the basic procedure given in Section 4.4 and modify this to produce procedures for the following:

   – arguments of type **bit** and std_logic

   – arguments of type **bit_vector** and std_logic_vector (these are registers)

   – use of reset and clear functions for different types of arguments

Step 2.   Analyze and test each of the procedures separately before committing them to placement within the package.

Step 3.   Define a new type designed to represent a 32-bit register. This is simply a 32-bit object of the type std_logic_vector:

> **type** register32 **is** std_logic_vector (31 **downto** 0);

Step 4.   Propose and implement one or two other types of objects that you may expect to find in a model of a CPU or memory system.

Step 5.   Create a library named MYLIB. This operation is usually simulator specific.

Step 6.   Compile the package into the library MYLIB. Your CAD tool documentation should provide guidelines on compiling design units into a library.

Step 7.   Write a VHDL model of a bit-serial adder, using signals of type **bit** and adopting the structure shown in Figure 6.10. The model must declare the library MYLIB and provide access to your package via the **use** clause.

Step 8.   Test the model of the bit-serial adder to ensure that it is functioning correctly.

Step 9. Modify the model to use signals of type std_logic. Nothing else should have to change, including the structure of the procedure call. By virtue of the argument type in the procedure call, the correct procedure in the package that you have written will be used.

Step 10. Modify the **use** clause to limit the visibility to one procedure in the package. Repeat your simulation experiments. You might have multiple instances of the **use** clause to provide visibility to each of the procedures you wish to utilize.

Step 11. Repeat the experiment to use other models of the D flip-flop that include signals such as reset and clear.

End Simulation Exercise 6.1

6.6   Chapter Summary

Designs can become large and complex. We need constructs that can help us manage this complexity and enhance the sharing of common design units. This chapter has addressed the essential issues governing the construction and use of subprograms: functions and procedures. We can organize commonly used subprograms into packages and place them in design libraries for subsequent reuse and sharing across distinct VHDL models. The concepts introduced in this chapter include the following:

- Functions
  - type conversion functions
  - resolution functions

- Procedures
  - concurrent procedure calls
  - sequential procedure calls

- Subprogram overloading
  - subprogram name
  - operator overloading

- Visibility rules
- Packages
  - package declaration
  - package body

- Libraries
  - relationships between design units and libraries
  - managing the scope and visibility of package contents

We are now armed with constructs for hiding complexity and sharing and reusing VHDL code modules.

Exercises

1. Create a package with functions or procedures for performing various shift, increment, and decrement operations on **bit_vector** elements and std_logic_vector elements. Place the package declaration and package body in distinct files and analyze them separately. Remember to declare and use the library IEEE and the package std_logic_1164.vhd.

2. Consider a VHDL type that can take on the values (0, 1, X, U). The values X and U correspond to the values unknown and uninitialized, respectively. Define a resolved type that takes on these values, and write and test a resolution function for this resolved type.

3. Write and test a set of procedures for performing arithmetic left and right shifts on vectors of type std_logic_vector.

4. Write and test a resolution function that operates on elements of type std_logic_vector and returns the largest value.

5. Write a function to compute the parity of a std_logic_vector. The value of the parity bit is set if the number of 1's in the value of data is odd. Use the exclusive-OR operation. Verify the correctness of the function via simulation.

6. Write and test functions that can perform type conversion between multibit quantities of type std_logic_vector and integers.

7. Using a concurrent procedure looks very much like using a component in a hierarchically structured design. What is the difference between using a concurrent procedure and constructing a structural design?

8. Create a design library My_Lib and place a package in this library. You might create a package of your own or simply "borrow" any one of a number of existing packages that come with VHDL environments. Analyze the package into this library. The creation of the library with the logical name My_Lib will involve simulator-specific operations. Ensure that you have correctly implemented the library by using elements of this package in a VHDL model analyzed into your working library, WORK.

CHAPTER 7

Basic Input/Output

Thus far, we have written VHDL programs that manipulate three classes of programming objects: variables, signals, and constants. We adhere to certain rules when using these objects. For example, if we were to use a variable in our program, we would give it a name such as Index, and we first declare the type of Index using a declaration statement. The range of values that the program can validly assign to Index, and the operations that can be performed on Index, are determined by its type. For example, if Index is an integer, we may perform integer arithmetic using Index, and the range of values that it may take depends on the number of bits used to represent an integer: 16, 32, and so forth. In an analogous fashion, the use of input/output functions necessitates the introduction of the *file* type that permits us to declare file objects.

Files are special and serve as the interface between the VHDL programs and the host environment. They are manipulated in a manner very different from variables or signals; hence there is a need for a distinct object type. As you might expect, there are special operations that are performed only on files: reading and writing. This chapter discusses how file objects are created, read, written, and used within VHDL simulations. A very useful example of the application of file I/O is the construction of testbenches—VHDL programs for testing VHDL models. The testbench reads test inputs from a file, applies them to the VHDL model under test, and records model outputs for analysis. The notion of a testbench follows directly from the testing of chips and boards and provides a structured approach for validating designs captured in VHDL simulation models.

7.1   Basic Input/Output Operations

As with variable, signal, and constant objects, before we can use a file object we must give it a name and declare its type. What determines the type of a file? A natural expectation is that the information provided in the file determines its type. For example, if a file contains a sequence of integers, we might naturally think of this file as being of type integer. As users, we tend to create files with all types of information—for example, strings, real numbers, and std_logic_vectors. To create and access information from these files, we would like a way to state the type of information contained in them and have functions for reading and writing that information. This chapter focuses on file input/output mechanisms in VHDL'93. File input/output in VHDL'87 is quite different and is explicitly marked in the text.

On the basis of our experiences with input/output in conventional programming languages and the preceding discussion, we can identify the following basic operations that we need for reading and writing files:

- Declaration of a file and its type
- Opening and closing a file of a specified type
- Reading and writing from a file

We now consider each of these steps and conclude this section with some examples.

7.1.1   File Declarations

If we want to declare a file, we also wish to make sure that the host environment can correctly interpret the format of the data stored in the file. This is achieved by first declaring the type of a file as follows:

> **type** TEXT **is file of string**;
> **type** INTF **is file of integer**;

The first declaration defines a *file type* that can store ASCII data. Files of this type contain human-readable text. The second declaration defines a file type that can store a sequence of integers. Files of this type are stored in binary form and are machine readable, but not human readable. Both **string** and **integer** are predefined types of the language, and their definition appears in the package STANDARD. In general, file types can describe sequences of more complicated data structures; however, we restrict ourselves to the essentials of the language and therefore the relatively simpler cases of predefined data types.

Now we can declare a file object as being of a particular file type as follows:

> **file** Integer_File : INTF;
> **file** Input_File : TEXT;

Note that the file type TEXT is also a predefined type in the package TEXTIO, which is typically distributed with all VHDL environments and is described in Section 7.2. We can think of Integer_File and Input_File as pointers to files that contain sequences of integers and characters, respectively. When statements in VHDL procedures read and write from these files, the file type enables the procedures to correctly interpret the values being read from the files.

7.1.2 Opening and Closing Files

Once we have declared files of a specific type, we must open them prior to using them and close them prior to termination of the program. In conventional programming languages, I/O operations involve calls to initialization procedures to open files *prior* to reading or writing data. Other procedures must also be called prior to closing files. For example, the C language provides the fopen() function call, which returns a pointer to a file. The fclose() function call closes the file and makes sure that the most recent updates have been written out to the file on disk (since they may still be cached in memory). In between calls to these two procedures, a file can be read or written. In VHDL'93, the following procedures open and close files.

> **procedure** FILE_OPEN (**file** file_handle : FILE_TYPE;
> File_Name : **in** STRING;
> Open_Kind: **in** FILE_OPEN_KIND := READ_MODE);
> **procedure** FILE_OPEN (File_Status: **out** FILE_OPEN_STATUS;
> **file** file_handle : FILE_TYPE;
> File_Name : **in** STRING;
> Open_Kind : **in** FILE_OPEN_KIND := READ_MODE);
> **procedure** FILE_CLOSE (**file** f : FILE_TYPE);

Consider the first procedure. The first argument is the file pointer that the VHDL program uses—for example, in the read and write procedures to be discussed in Section 7.1.3. The second argument is the name of the file that will be read or written. For example, you might have a file named *input_file.txt* in your working directory. The third argument describes how the file is to be used—that is, the mode of the file. A file can be opened in three modes—READ_MODE, WRITE_MODE, and APPEND_MODE. The default mode is READ_MODE. The second procedure is the same as the first, with the addition of a new parameter, File_Status. This parameter has a value returned by the procedure that may have one of four values:

OPEN_OK	file open operation was successful
STATUS_ERROR	attempted to open an already open file
NAME_ERROR	file not found
MODE_ERROR	file cannot be opened in this mode

The File_Status variable is an enumerated type whose value can be checked to make sure that the FILE_OPEN call was successful. Finally the FILE_CLOSE procedure closes the file. FILE_CLOSE is implicitly called when execution terminates; thus, you do not have to explicitly call FILE_CLOSE. As a matter of fact, we can even avoid calling FILE_OPEN by providing the same information in the file declaration as is illustrated in the next example.

Example: Explicit vs. Implicit File Open

The following template of code is typically found in processes that use files (in the file open call, the name of the file can include a path in your local machine hierarchy):

```
type IntegerFileType is file of integer; -- declare a file type in the
                                       -- -- architecture declarative region
process is - a template for a process
file datain :IntegerFileType; -- declare file handle
variable fstatus :File_open_status; -- declare file status variable
--
-- other misc declarations
--
begin
file_open(fstatus, dataout,"myfile.txt", read_mode);
--
-- body of process: reading and writing files and
-- performing computations
--
end process; -- termination implicitly causes a call to FILE_CLOSE
```

Files can be opened implicitly by providing the necessary information in the file declaration as follows (there is no explicit FILE_OPEN call):

```
type IntegerFileType is file of integer; -- declare a file type in the
                                       -- -- architecture declarative region
process is - a template for a process
file datain:IntegerFileType open read_mode is "myfile.txt"; -- declare file
                                                             handle
--
-- other misc declarations
--
begin
--
-- body of process: reading and writing files and
-- performing computations
--
end process; -- termination implicitly causes a call to FILE_CLOSE
```

Example End: Explicit vs. Implicit File Open

7.1.3   Reading and Writing Files

Now that we can declare files of a certain type and open and close these files prior to and after their use, we need procedures and functions to read and write files of the declared type. What are these I/O procedures and how are they used? The arguments of these functions include a pointer to the file and the variables to be written or read. Most other programming languages provide equivalent functions to be used in a similar manner. In principle, the VHDL language operates in much the same way, with a few important differences. The standard VHDL procedures made available by the definition of the language are as follows:

> **procedure** READ (**file** file_handle : FILE_TYPE; value : **out** type);
> **procedure** WRITE (**file** file_handle : FILE_TYPE; value : **in** type);
> **function** ENDFILE (**file** file_handle : FILE_TYPE) **return boolean**;

These I/O procedures are implicitly declared following a file type declaration. This means that you do not have to declare these procedures prior to their use. Whereas the procedures READ and WRITE apply to input/output operations, the ENDFILE function tests for the end of file when reading from files. Most VHDL simulators will support a set of procedures for reading and writing various data types: **character**, **integer**, **real**, and so on. (Refer to your simulator documentation for more details on the available I/O procedures for reading and writing different data types.) The standard package TEXTIO provides procedures for reading and writing files of type TEXT. More information on the package TEXTIO is provided in Section 7.2.

While this text focuses on the basic features of the language advanced topics covering input/output can be found in several excellent texts [5,14].

'87 vs. '93
☞ ### 7.1.4   VHDL 1987 I/O

File type declarations within VHDL'87 are the same as in VHDL'93. However, the file declarations, which provide pointers to files, are different and incorporate the functionality of the FILE_OPEN procedure in VHDL'93. The VHDL'87 file declarations will appear as follows:

> **File** infile : intf **is in** "inputdata.txt";
> **File** outfile : INTF **is out** "outputdata.txt";

Note that the file type INTF : VHDL is case insensitive, so both statements refer to the same filetype. The object infile can be regarded as a pointer to a file to be opened for input. The name of the file is inputdata.txt. Since the file is open for input, the access mode of the file is said to be of type **in**. The default location of this file is the current working directory. Similarly, the object outfile is declared as a pointer to a file named outputdata.txt. The access mode for this latter file is declared to be of type **out** and can only be written. When VHDL was revised in 1993, changes were introduced in the input and output operations. In contrast to VHDL'87, file declarations in VHDL'93 appear as follows:

> **File** infile : text **open** read_mode **is** "inputdata.txt";
> **File** outfile : text **open** write_mode **is** "outputdata.txt";

The file objects infile and outfile can be used by VHDL procedures to read and write the files inputdata.txt and outputdata.txt, respectively, using the **read**() and **write**() procedures as defined in Section 7.1.3:

> **procedure** READ (**file** file_handle : FILE_TYPE; value : **out** type);
> **procedure** WRITE (**file** file_handle : FILE_TYPE; value : **in** type);
> **function** ENDFILE (**file** file_handle : FILE_TYPE) **return boolean**;

We see that the syntax of these procedures is essentially the same in VHDL'93. Finally, note that in VHDL'87 there are no explicit FILE_OPEN and FILE_CLOSE procedures. These operations are performed implicitly during the file declarations, as shown earlier in this section.

Example: Basic Binary File I/O

The code in Figure 7.1 shows an example of basic binary file output. We wish to write a file with integer values. Therefore, we declare a file type IntegerFileType and a file pointer dataout as a pointer to a file of that type. Now we are ready to write to this file.

```
entity io93 is -- this entity is empty
end entity io93;

architecture behavioral of io93 is
begin
process is
type IntegerFileType is file of integer; -- file declarations
file dataout :IntegerFileType;
variable count : integer:= 0;
variable fstatus: FILE_OPEN_STATUS;

begin
file_open(fstatus, dataout,"myfile.txt", write_mode); -- open the file
for j in 1 to 8 loop
write(dataout,count); -- some random values to write to the file
count := count+2;
end loop;
wait; -- an artificial way to stop the process
end process;
end architecture behavioral;
```

FIGURE 7.1   Basic binary file input/output

The file is initialized in the body of the process by opening the file "myfile.txt" in mode write_mode. The text following the file open procedure is a simple loop that generates a sequence of integer values to be written to the file. The code block ends rather artificially, with the process suspended by the **wait** statement. To read the file "myfile.txt", we can imagine another VHDL model with a block of code that looks exactly like the preceding **for-loop**, but with the write(dataout, count) procedure replaced by a read(dataout, myinteger[i]), so that the sequence of values can be read into an integer array, myinteger[]. The mode of the FILE_OPEN() procedure would also have to be read_mode rather than write_mode.

Example End: Basic Binary File I/O

The language supports reading and writing from files of the predefined types of the language. If you wish to read or write other types not defined by the language, you must write your own input/output procedures built on these basic procedures. It is common to do so and to encapsulate these procedures and any new type definitions in a package. In fact vendors will often supply a set of procedures and functions for reading and writing specific data types. Check the vendor documentation for more specific information.

One common approach to input/output for the predefined types of the language is the use of the TEXTIO package, described in the next section.

7.2   The Package TEXTIO

The package TEXTIO is a standard package supported by all VHDL simulators. Recall that a package can be thought of as a library or repository of predefined types, variables, signals, constants, functions, and procedures. The TEXTIO package provides a standard set of file types, data types, and input/output functions. In general, when you use a package in a VHDL model, you must declare the package to be used. The declaration will state the library (system directory) that serves as the location of the package. The TEXTIO package is in the library STD, which is implicitly declared. This means that you do not have to declare the use of STD, as we do with the library IEEE. However, you must declare the usage of the package contents via the **use** clause as shown in the examples that follow.

The package TEXTIO defines a standard file type called TEXT and provides the procedures for reading and writing the predefined types of the language, such as **bit**, **integer**, and **character**. These are shown in the package definition given in Appendix B (including the text object type and file handles for std_input and std_output). From Appendix B, we see that **read** () and **write** () procedures are defined for several data types. Normally, we expect that each data type will have a distinct procedure for reading and writing the type to a file. For example, we could have read_string () and read_bit_vector (). However, given the large number of data types, this quickly

becomes a rather tedious exercise in naming procedures. Instead, it is convenient for
all of the procedures that perform the same function, such as **read**(), to have the same
name, and permit the actual argument to identify the correct implementation: **read** ()
and **write** () are overloaded procedure names. We do not have to remember the exact
name of the procedure to read and write elements of type **bit_vector** or **string**. We simply use the procedures **read** () and **write** (), and, depending on whether the argument
is a **bit_vector** or **string**, the appropriate implementation is invoked. We see that **read** ()
and **write** () procedures are available for the predefined types **bit**, **bit_vector**, **character**, and **string**.

The use of the TEXTIO package is based on the view of input/output operations
described shortly. Imagine a buffer that exists between the file you wish to read and
the VHDL program. Think of this buffer as a type of "staging area," and in our discussion let us refer to the buffer as buf. The **read**() and **write**() procedures access and
operate on buf, reading the values of program variables from buf or writing the values
of program variables into buf. The **readline**() and **writeline**() procedures move the contents of buf to and from files, respectively. These procedures are defined in the package
TEXTIO that is contained in the library STD. Two "special" buffers called input and
output are predefined in the package TEXTIO. These buffers are mapped to the
std_input and std_output of the host environment, which usually is the console window of the VHDL simulator. This reading from and writing to input and output,
respectively, will read from and print to the simulator console window. Let us look at
an example of the use of the TEXTIO package.

Example: Use of the TEXTIO Package

Figure 7.2 illustrates an example of the use of the TEXTIO package. The pointer outfile
is declared to point to a file of type text. Again, note that VHDL is not case sensitive,
and therefore text, Text, and TEXT are all identical. Statement L1 writes the text string
"This is an example of formatted IO" to the buffer buf. Statement L2 then writes the
buffer to the file, causing this text string to appear on a line in the output file. The next
sequence of write operations in statements L3 through L7 is written in the buffer and
when statement L8 is executed, the contents of writes in L3 through L7 will appear on
the same line. Viewed with a text editor, the output file will appear as follows:

> This is an example of formatted IO
> The First Parameter is = 5 The Second Parameter is = 0110
> ...and so on

A few items are worth noting. The variable buf can be accessed only via the **read** ()
and **write** () procedures. This is because buf is of type line, which is a special type
referred to as an **access** type. An access type is similar to a pointer in C or Pascal.
While access types provide powerful programming flexibility, we are more concerned
here with the hardware modeling aspects of VHDL. Therefore, other than its use in
file I/O, we will not deal with access types any further. Statements L4 and L7 each
write a distinct data type to buf. The **write** procedures are overloaded (as are the **read**

```
use STD.Textio.all;
entity formatted_io is -- this entity is empty
end formatted_io;

architecture behavioral of formatted_io is
begin

process is
file outfile :text; -- declare the file to be a text file
variable fstatus :File_open_status;
variable count: integer := 5;
variable value : bit_vector(3 downto 0):= X"6";
variable buf: line; -- this is the buffer between the program and the file

begin
file_open(fstatus, outfile,"myfile.txt", write_mode); -- open the file for writing
L1: write(buf, "This is an example of formatted IO");
L2: writeline(outfile, buf); -- this procedure writes the buffer to a line in the file
L3: write(buf, "The First Parameter is =");
L4: write(buf, count); -- this procedure writes to the next location in the buffer
L5: write(buf, ' ')L6: write(buf, "The Second Parameter is =");
L7: write(buf, value);
L8: writeline(outfile, buf);
L9: write(buf, "...and so on");
L10: writeline(outfile, buf);
L11: file_close(outfile); -- flush the buffer to the file
wait;
end process;
end architecture behavioral;
```

FIGURE 7.2   File I/O with the TEXTIO package

procedures) and support the predefined types. Writing **bit_vector** and **integer** types with the TEXTIO package creates files that are human readable. The file_close() procedure call will ensure that all writes to a file are complete. In general, it is not necessary to call this procedure explicitly; the procedure will be called implicitly when the process terminates. However, the process in Figure 7.2 suspends execution indefinitely on the **wait** statement. Again, this is an artifact of the contrived nature of the example. In general, if you know that your simulation will terminate, file_close() will be called on program termination.

Finally, if we were to replace the variable outfile with the variable output in all of the **writeline**() procedure calls, then the text output would be printed to std_output, rather than to the file *myfile.txt*. Usually, std_output is the simulator console.

Example End: Use of the TEXTIO Package

The preceding example uses the **write**() procedures on the predefined types of the language. What about types such as std_logic_vectors? To use the TEXTIO package with any of these types we need to convert them to characters for writing into the buffer buf. The next example illustrates the use of writing and the use of type conversion functions in a package.

Example: Basic Text Input/Output Operations

Suppose that we wish to read and write signal values of the type std_logic_vector. This example illustrates how we can develop input/output procedures to read and write signal values based on the available elementary character I/O operations that are provided by VHDL. If we can provide these new procedures in a package, we can then use these procedures in our VHDL models and hide the fact that we are using character I/O. This is exactly what most VHDL environments do by providing a set of functions in the package TEXTIO to read various data types. Note that we do need the **use** clause, so that the contents of the TEXTIO package are visible to the VHDL programs.

Figure 7.3 shows a package with two procedures. The first procedure is for reading a bit vector of type std_logic_vector and the second is for writing a bit vector of type std_logic_vector. Each procedure requires a file pointer as an argument. Each

```
library IEEE;
use IEEE.std_logic_1164.all;
use STD.textio.all;

package classio is
procedure read_v1d (variable f: in text; v : out std_logic_vector);
procedure write_v1d (variable f:out text; v : in std_logic_vector);
end package classio;

package body classio is
procedure read_v1d (variable f:in text; v : out std_logic_vector) is
variable buf: line;
variable c : character;

begin
readline(f, buf);
for i in v'range loop
read(buf, c);
case c is
 when 'X' => v (i) := 'X';
 when 'U' => v (i) := 'U';
 when 'Z' => v (i) := 'Z';
 when '0' => v (i) := '0';
 when '1' => v (i) := '1';
```

FIGURE 7.3 An example of using character I/O to read and write single-bit vectors

```
when '-' => v (i) := '-';
when 'W' => v (i) := 'W';
when 'L' => v (i) := 'L';
when 'H' => v (i) := 'H';
 when others => v (i) := '0';
end case;
end loop;
end procedure read_v1d;

procedure write_v1d (variable f: out text; v : in std_logic_vector) is
variable buf: line;
variable c : character;

begin
for i in v'range loop
case v(i) is

when 'X' => write(buf, 'X');
 when 'U' => write(buf, 'U');
 when 'Z' => write(buf, 'Z');
 when '0' => write(buf, 'character'('0'));
 when '1' => write(buf, 'character'('1'));
   when '-' => write(buf, '-');
 when 'W' => write(buf, 'W');
 when 'L' => write(buf, 'L');
 when 'H' => write(buf, 'H');
 when others => write(buf, 'character'('0'));
end case;
end loop;
writeline (f, buf);
end procedure write_v1d;
end package body classio;
```

FIGURE 7.3   (Continued)

procedure also accepts as a parameter a variable v of type std_logic_vector. We can use the same procedures for reading and writing bit vectors of differing precision. Note how each procedure works. In the read procedure, a complete line is read into the buffer buf, using the **readline**() VHDL procedure call. Once the complete vector is read into buf, this buffer is scanned to test for the value of each digit in the vector and to set the value of each output digit accordingly, using the **case** statement. Finally, note that the loop scanning the input vector executes a number of times that is determined by the **'range** attribute of the vector being read. An analogous procedure writes bit vectors of type std_logic_vector out to a file. Since we are dealing with text-based I/O in this example, in writing the bit values 0 and 1, we must first convert them to characters, as shown in write_v1d() in the figure. This is necessary so that the implementation can distinguish between the values of 0 and 1, on the one hand, and their corresponding ASCII character representations on the other. By using the basic read and write procedures provided by VHDL, we are now able to read and write single-bit vectors of the type std_logic_vector. This package can be compiled into a library or simply into the local working directory that is accessed via the logical name WORK.

Example End: Basic Text Input/Output Operations

We can similarly construct procedures for reading and writing other data types from files. Generally, rather than having the user develop these procedures, the CAD tool vendor provides a comprehensive set of input/output functions for reading and writing various data types.

The package described in Figure 7.3 can be used within a VHDL model for reading std_logic_vector values from a file and using them internally, or for writing generated std_logic_vectors to a file for later analysis. A simple template for using the package *classio.vhd* is described in the next example.

Example: Using the Classio Package

Figure 7.4 shows an example of how the procedures and the package shown in Figure 7.3 might be used. The code simply reads 16-bit vectors from infile.txt and writes them out to the file outfile.txt. Note that, by changing the resolution of the variable check to 32 bits, we can use the same procedures for reading and writing 32-bit vectors. This is because the loop in the read/write procedures that scans the vectors is parameterized by the range of the vector, as opposed to having a fixed value. The simple test program uses a **wait** statement to ensure that the simulation progresses. Remember that, in general, processes are executed once, when the simulation is initialized. Thereafter, processes are executed only if they are invoked by events, which may occur in a sensitivity list or via the use of **wait** statements. In general, you will probably be performing I/O operations within some process that will be invoked due to some simulation events; therefore, the use of the **wait** statement in the preceding example should be regarded as somewhat artificial.

Example End: Using the Classio Package

We have now covered the essentials of binary and text file input/output. The next section focuses on the use of file input/output as part of the important process of testing and validation of VHDL models.

```
library IEEE;
use IEEE.std_logic_1164.all;
use STD.textio.all;
use Work.classio.all; -- the package classio has been compiled into
                      -- the working directory
entity checking is
end entity checking; -- the entity is an empty entity

architecture behavioral of checking is
begin
process is
-- use implicit file open
--
file infile : TEXT open read_mode is "infile.txt";
file outfile : TEXT open write_mode is "outfile.txt";
variable check : std_logic_vector (15 downto 0) := x"0008";

begin
-- copy the input file contents to the output file
while not (endfile (infile)) loop
read_v1d (infile, check);
write_v1d (outfile, check);
end loop;
file_close(outfile); -- flush buffers to output file
wait; -- artificial wait for this example
end process;
end architecture behavioral;
```

FIGURE 7.4   An example of the use of the package *classio.vhd*

Simulation Exercise 7.1: Basic Input/Output

This exercise introduces the student to the basic steps involved in reading and writing text files and covers the following concepts: (i) declaring and opening files, (ii) initializing data structures from files, and (iii) recording simulation results in files.

Step 1. Create a text file, *memory.vhd*, with the model of a memory module from Figure 4.2. The succeeding steps will modify this model to initialize the memory to encoded values read from a file.

Step 2. Create the file *classio.vhd* shown in Figure 7.3. This is a package. Compile this package into your working directory. The simulator you are using should have the default library WORK set to this directory.

Step 3. Edit *memory.vhd* to enable references to the package *classio.vhd* by placing the following clause at the beginning of the file:

use WORK.classio.all;

Step 4. Create a text file *infile.txt*. This file will include the contents of memory. Each line should be a 32-bit vector of type g, and there should one word per line. For the model we are using, the file should have eight words.

Step 5. Add a single-bit signal, reset, as an input signal to the entity description. This signal is of mode **in** and of type std_logic. A pulse on this signal will indicate that the memory contents are to be initialized from a file.

Step 6. Add a separate process to the model to read eight 32-bit values from a file and initialize the contents of memory to these values.

> **Step 6(a)** Label this process ioproc.
>
> **Step 6(b)** Have the first line in the process be a **wait** statement, causing the process to wait for a rising edge on the signal reset. Examples of the detection of rising edges on signals can be found in the D flip-flop examples in Chapter 4.
>
> **Step 6(c)** Use the functions in the package classio in writing the body of the process. The process body reads and initializes the contents of memory from the input file. The model in Figure 7.4 illustrates how a **while** loop can be used for this purpose.

Step 7. The modified memory model is now complete. Compile the model into the working directory.

Step 8. Test the model as follows:

> **Step 8(a)** Create a sequence of test inputs to the model. For the reset signal, we must supply a single pulse. For the remaining inputs, supply a sequence of addresses starting from 0 and proceeding through memory address 7. The memory read control signal must be asserted.
>
> **Step 8(b)** Select the signals to be traced. Trace all of the signals in the entity description.
>
> **Step 8(c)** Apply the stimulus as generated in Step 8(a). Observe the trace.
>
> **Step 8(d)** After eight memory read operations, the values read from memory should be identical to the values initialized from the file.

Step 9. Modify the model to use an output file *outfile.txt*. Now generate a sequence of read and write operations to memory, using the external stimulus capabilities of your simulator.

Step 10. Now modify the model to log all memory operations to the output file. For each read and write operation, the model records the value and address in *outfile.txt*.

Step 11. Verify that the contents of *outfile.txt* agree with the test sequence that you generated in Step 9.

Step 12. Modify the memory model so that data types used are **bit** and **bit_vector** rather than std_logic and std_logic_vector. For example, memory would be an array of words of type **bit_vector**. Now we can use the I/O functions in the package TEXTIO. These functions are shown in Appendix B. Modify the model to use the I/O functions from this package.

Step 13. Rerun the model by using the TEXTIO package rather than the classio package. I/O procedures are available for all of the predefined types of the language—for example, strings. Modify the **use** clause accordingly. Remember, std_logic and std_logic_vector are not predefined types of the language. They are defined by the package std_logic_1164!

End Simulation Exercise 7.1

The examples we have seen up to this point have had filenames specified in the file declarations or file open procedures. As with conventional languages, it is desirable to be able to specify the filename and thereby avoid recompiling the simulation model each time we change the input file.

Example: Filenames as Inputs

We would like to be able to read a character string corresponding to the filename from the simulator console. Fortunately for us, the TEXTIO package defines file handles for std_input and std_output as input and output, respectively. The following process code block can be used to read a file name from the simulator console, assuming that the simulator console is mapped to std_input:

```
process is
variable buf : line;
variable fname : string(1 to 10);
begin
--
-- prompt and read filename from standard input
--
write(output, "Enter Filename: ");
readline(input, buf);
read(buf, fname);
--
-- process code
--
end process;
```

The variable fname can now be used in the file_open procedure.

Example End: Filenames as Inputs

Now that we have the ability to perform file input/output, what else can we do, other than the obvious? The next section describes a very important process central to digital design, namely, the testing and validation of VHDL models. File input/output is routinely used in this process.

7.3   Testbenches in VHDL

Much of the motivation for simulation is to be able to test designs prior to construction and use of the circuit. To motivate a testing methodology, we might consider a physical analogy. How would we test an electronic component, such as a silicon chip, that implements some digital circuit? Intuitively, we would like to provide a set of inputs and check the observed output values against the corresponding correct set of output values. The chip can be placed in a specialized piece of equipment, referred to as a test frame, that allows us to apply an electrical stimulus to the inputs and examine the values of the outputs with a logic analyzer. By cycling through possible input sequences, we could analyze the corresponding output sequences to determine whether the component was functioning correctly.

In developing VHDL models, we find ourselves in a similar situation. We construct a VHDL simulation model of some digital system component, such as an encoder for audio signals. How do we test this model to ensure that (i) the model is operating as designed and (ii) the design itself is correct? Verifying a design via simulation is certainly more cost effective than testing a fabricated part and then determining that it has a design error that must be fixed. If our VHDL model is sufficiently detailed, then thorough testing in simulation significantly reduces the chances of errors, minimizes costly design rework, and reduces the time to get the product to the marketplace.

Most simulators provide commands to apply stimulus to the input ports of a design entity. By tracing and viewing the resulting values of the signals on the output ports, we can determine whether the model is operating correctly. However, by recognizing that VHDL provides powerful programming language abstractions for describing the operation of digital systems, we can realize a more structured approach toward testing VHDL models. The test frame described for placing and testing chips is itself a digital system and we should be able to describe the operation of the tester in VHDL! The logical behavior of the tester is simple to understand. The model generates sequences of inputs and reads the outputs of the module being tested. This behavior is captured in the notion of a *testbench* and is illustrated in Figure 7.5.

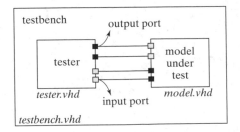

FIGURE 7.5   The structure of a testbench

In the figure, one VHDL module (*tester.vhd*) generates the stimulus to be applied. A second VHDL module (*model.vhd*) is the model being tested. Finally, a third module (*testbench.vhd*) is a structural VHDL model that describes the interconnections between the tester and the model under test. This model describes how the ports of the tester are connected to ports of the model. The simulation progresses with the tester applying a stimulus to the model and reading and recording the responses. The sequences of input values to be applied to the input ports of the model under test can be read from a file, or they can be generated in VHDL. For example, clock signals, reset pulses, and other types of periodic waveforms can be generated using approaches described in Section 4.6. The tester then checks the results returned from the VHDL model against a known correct set of output values and flags any errors that are found. In testing large systems with a potential of millions of test vectors, automation of this process is a very productive enterprise. The construction of testbenches is best illustrated with an example.

Example: Writing a Testbench

Let us assume that we wish to write a testbench to test the VHDL model of the positive edge-triggered D flip-flop developed in Chapter 4 and shown in Figure 7.6. In

```vhdl
library IEEE;
use IEEE.std_logic_1164.all;

entity asynch_dff is
port (R, S, D, Clk : in std_logic;
      Q, Qbar : out std_logic);
end entity asynch_dff;

architecture behavioral of asynch_dff is
 begin
output: process (R, S, Clk) is
 begin
if (R = '0') then
    Q <= '0' after 5 ns;
    Qbar <= '1' after 5 ns;
 elsif S = '0' then
    Q <= '1' after 5 ns;
    Qbar <= '0' after 5 ns;
  elsif (rising_edge(Clk)) then
    Q <= D after 5 ns;
    Qbar <= ( not D) after 5 ns;
 end if;
 end process output;
 end architecture behavioral;
```

FIGURE 7.6   Behavioral model of a positive edge-triggered D flip-flop

order to test this model, we must generate a clock signal and a sequence of values on the D input signals. We must also test the Set (S) and Reset (R) operations. Figure 7.7 shows a sample test pattern for the clock and input (D). The D input is sampled at the rising edge of the clock, producing the output waveforms shown in the figure. The Set (S) input and Reset (R) inputs are tested by applying a test vector where either S or R or both are asserted and then reading the value of the flip-flop outputs after a delay equal to the propagation delays of the signals through the flip-flop.

Figure 7.8 shows an example of a tester module that generates the clock signal and applies a set of test vectors to the D flip-flop model. Note that the sample I/O package shown in Figure 7.3 is compiled into the library WORK. Consider the structure of the tester module. The process named clk_process generates a clock signal with a period of 20 ns. This process executes concurrently with the io_process. This latter process reads test vectors from file *infile.txt*. Each test vector is five bits long. The first three bits correspond to input values for R, S, and D. The last two bits are the corresponding correct values of Q and Qbar. Collectively, these five values constitute one test vector. The input test vector values are applied to the flip-flop model at 20-ns intervals—the rising edge of the clock. From the correct timing behavior shown in Figure 7.7, we can generate a set of test vectors by looking at the values of the inputs on each rising edge (i.e., at time 20 ns, 40 ns, and so on). This produces the following test vectors in file *infile.txt*:

```
11001 -- initial vector
01101
11110
10010
10011 -- invalid case
```

Each vector is applied at 20-ns intervals to coincide with the rising edge of the clock. The last test vector corresponds to an invalid case, since both Q and Qbar cannot be asserted at the same time. Therefore, this test vector will fail and an error message will be logged to the output file *outfile.txt*. This example also shows that the

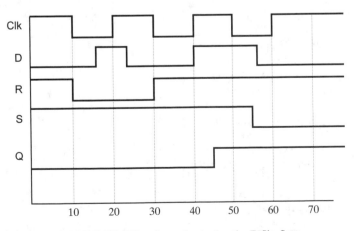

FIGURE 7.7   Waveforms for testing the D flip-flop

```
library IEEE;
use IEEE.std_logic_1164.all;
use STD.textio.all;
use WORK.classio.all; -- declare the I/O package

entity srtester is                          -- this is the module generating the tests
port (R, S, D, Clk : out std_logic;
      Q, Qbar : in std_logic);
end entity srtester;

architecture behavioral of srtester is
begin
clk_process: process   is -- generates the clock waveform with
begin                                -- period of 20 ns
Clk<= '1', '0' after 10 ns, '1' after 20 ns, '0' after 30 ns;
wait for 40 ns;
end process clk_process;

io_process: process is -- this process performs the test
file infile : TEXT open read_mode is "infile.txt";          -- functions
file outfile : TEXT open write_mode is "outfile.txt";
variable buf : line;
variable msg : string(1 to 20) := "This vector failed!";
variable check : std_logic_vector (4 downto 0);
begin
while not (endfile (infile)) loop -- loop through all test vectors in
read_v1d (infile, check);          -- the file
R <= check(4);
S <= check(3);
D <= check(2);
wait for 20 ns;                     -- wait for outputs to be available after applying
                                    -- the stimulus
if (Q /= check (1) or (Qbar /= check(0))) then -- error check
write (buf, msg);
writeline (outfile, buf);
write_v1d (outfile, check);
end if;
end loop;
file_close(outfile); -- flush contents to file
wait;      -- this wait statement is important to allow the simulation to halt!
end process io_process;
end architecture behavioral;
```

FIGURE 7.8 Behavioral description of the tester module

failure of a test vector may be due not to an incorrect model, but to an incorrect test vector. In fact, running these test vectors as given will also generate an error for the first vector, since at this time the values of Q and Qbar are undefined or U. We should build flip-flop models with global reset to a known state. The generation of test vectors in general is a computationally intensive and nontrivial task. Once the stimulus is applied, the process waits for 20 ns. By this time the flip-flop outputs become stable (see Figure 7.6). Also at this time, the outputs are read and compared with known correct values that were read from the input file. If there is any discrepancy in these values, the corresponding test vector is flagged by writing out an error message to an output file along with the test vector. After all the test vectors have been read and applied, the output file contains the list of test vectors for which the model failed.

The top-level module for our testbench is a structural model for specifying the connections between the tester and the model under test and is shown in Figure 7.9. Note the use of configuration specifications to explicitly state which VHDL architectures are to be used for the design entities srtester and asynch_dff. In general, more than one architecture could have been used. The configuration specification states that

```
library IEEE;
use IEEE.std_logic_1164.all;
use WORK.classio.all; -- declare the I/O package

entity srbench is    -- the entity interface is empty
end entity srbench;

architecture behavioral of srbench is
component asynch_dff is
port (R, S, D, Clk : in std_logic;
      Q, Qbar : out std_logic);
end component asynch_dff;

component srtester is
port (R, S, D, Clk : out std_logic;
      Q, Qbar : in std_logic);
end component srtester;
--
-- configuration specification
--
for T1:srtester use entity WORK.srtester (behavioral);
for M1: asynch_dff use entity WORK.asynch_dff (behavioral);

signal s_r, s_s, s_d, s_q, s_qb, s_clk : std_logic;

begin
T1: srtester port map (R=>s_r, S=>s_s, D=>s_d, Q=>s_q, Qbar=>s_qb,
                        Clk => s_clk);
M1: asynch_dff port map (R=>s_r, S=>s_s, D=>s_d, Q=>s_q,
                          Qbar=>s_qb, Clk => s_clk);
end architecture behavioral;
```

FIGURE 7.9   Structural description of the testbench module

the architecture labeled behavioral is to be used for each entity. The configuration also states that these architectures can be found in the working directory denoted by the library WORK. If no configuration specification had been provided, default rules would apply, and we would have used the last compiled architecture for entities srtester and asynch_dff.

In testing combinational circuits, that there may be no clock or periodic signal. In that case, we can write the tester module without having to be concerned about synchronizing with a periodic signal. The module can apply the input vectors, wait for a period equal to the propagation delay of the longest path through the circuit, and then read the output signal values of the module under test.

Example End: Writing a Testbench

7.4 ASSERT Statement

In the previous example, the testbench model recorded errors or failures to pass a test vector by writing the failed test vector and a brief message to the file outfile.txt. Alternatively, we can use the **assert** statement, which is a general mechanism for detecting and reporting incorrect conditions during a simulation. We can use this statement to detect failed test vectors, as shown in Figure 7.10. This statement would replace the **if** statement in the srtester module in Figure 7.8 that checks the output signal values and causes error messages to be written to a file. If we use the **assert** statement, rather than having an error message and the offending test vector being written to a file, the message regarding a violation and the string provided with the **report** clause would be sent to the simulation output. For example, Figure 7.10 shows the contents of the simulator console when executing the example testbench in Figure 7.9. This is the output from the Active VHDL simulator.

Note the absence of a semicolon after the report clause. This output is generally the simulator console window, unless you have redirected it to a file. The designer

```
assert Q = check(1) and Qbar = check(0)
report "Test Vector Failed"
severity error;
```

Example of Simulator Console Output

```
Selected Top-Level: srbench (behavioral)
: ERROR  : Test Vector Failed
: Time: 20 ns, Iteration: 0, Instance: /T1.
: ERROR : Test Vector Failed
: Time: 100 ns, Iteration: 0, Instance: /T1.
```

FIGURE 7.10   An example of the use of the **assert** statement in the testbench shown in Figure 7.8 and the resulting message printed in the simulator console

can report messages at one of several predefined severity levels: NOTE, WARNING, ERROR, and FAILURE. This provides a clean mechanism for the designer to classify the levels of information conveyed during simulation. For example, the NOTE category can provide information about the progress of the simulation, while a severity level of ERROR may cause the simulation to be aborted.

7.5   A Testbench Template

We are now ready to outline a few basic steps toward constructing a testbench for testing a VHDL model. It should be clear from the preceding examples that the testbench is a structural model with two components: a tester and the model under test. This is not the only way to structure a testbench, but it is intuitive, common, and simple to construct.

The model under test may be a behavioral or structural VHDL model of a digital system. The tester is usually a behavioral model, written using the constructs described in Chapter 4. Typical segments of VHDL code that we may find in tester modules include (i) processes to generate waveforms, (ii) VHDL statements to read test vectors from input files and apply them to the model under test, and (iii) VHDL statements to record the outputs that the model produces under test in response to the test vectors. A template for such a top-level structural model is shown in Figure 7.11. A step-by-step description for the construction of such a testbench model is the same as that for creating structural models as described in Chapter 5.

Recall from Chapter 4 that we can mix concurrent and sequential statements (via processes) in the architecture description of a circuit. Rather than have distinct models for the testbench module, tester module, and the model under test, the tester VHDL code may be directly included in the testbench architecture. The component instantiation statement T1 in Figure 7.9 can be replaced by the tester code. If we took this approach we would have only two code modules: the testbench and the model under test.

The models we are dealing with here are relatively small. We can see that as models become more complex, the number and size of the test vectors could become very large. For example, consider a circuit that processes 32-bit data according to a 16-bit operation code. At the very least, this circuit will have 48 inputs. A naive testing approach would attempt to test all possible input combinations. The total number of possible combinations of input values (assuming only 0 and 1 values) is 2^{48}! Even if we could perform each test in a nanosecond, it would still take on the order of thousands of years to finish the test suite. Furthermore, the number of inputs in modern chips and systems is considerably higher. In reality, the number of input combinations that must be tested can be pared considerably, and many computational as well as design techniques have been implemented to further reduce the cost of testing. Even so, it is apparent that the generation of test vectors can be a very complex process in its own right. To facilitate the sharing of test vectors among groups (e.g., the people generating them and the designers using them), standards are often defined for the specification of these vectors.

```
library Lib1;     -- declare any libraries that will be needed
library Lib2;
use Lib1.package_name.all; -- declare the packages that will be used
use Lib2.package_name.all; -- in these libraries
entity test_bench_name is
port( input signals : in type;
      output signals : out type);
end entity test_bench_name;

architecture arch_name of test_bench_name is

-- declare tester and model components
component tester_name is
port ( input signals : in type;
      output signals : out type);
end component tester_name;

component model_name is
port( input signals : in type;
      output signals : out type);
end component model_name;

-- declare all signals used to connect the extra tester and model

signal internal signals : type := initialization;

begin

-- label each component and connect its ports to signals or other ports

T1: tester-name port map (port=> signal1,.....);

M1: model-name port map (port => signal2,.....);

end architecture arch_name;
```

FIGURE 7.11 A testbench model template

Finally, remember that input or output functions are often expected to be executed only once, such as in loading memory in a processor datapath model. If the I/O code is placed in a part of a process that is executed repetitively, array-out-of-bounds errors can occur.

Simulation Exercise 7.2: Constructing Testbenches

This exercise familiarizes the student with the basic steps involved in constructing and executing a testbench for testing and validating a VHDL model. Basic concepts covered include (i) generating a stimulus for a model under test, (ii) reading test vectors from a file and applying them to the VHDL model under test, and (iii) recording failed test vectors and generating error messages for examination.

Step 1. Using a text editor, create the files *dff.vhd*, *srtester.vhd*, and *testbench.vhd* shown in Figure 7.6, Figure 7.8, and Figure 7.9, respectively. These files should be in your working directory. Compile each file.

Step 2. Create the file *classio.vhd* shown in Figure 7.3. This is a package. Compile this package into your working directory. The simulator you are using should have the default library WORK set to this directory.

Step 3. Create a text file named *infile.txt*. This file's contents should appear as follows:

```
11001
01101
11110
10010
```

This file represents a set of test vectors to be applied to the model *dff.vhd*. The first three bits (left to right) represent test inputs for the R, S, and D inputs, respectively. The last two inputs represent the corresponding correct values for Q and Qbar, respectively.

Step 4. Create the empty text file *outfile.txt*.

Step 5. Run the simulation long enough to apply the test vectors.

Step 6. Examine the output file *outfile.txt* for any error messages. How many error messages should appear, if any, for the preceding sequence of test vectors?

Step 7. Modify some of the test vectors in *infile.txt*. Rerun the simulation and study the input and output files to determine whether the model is functioning correctly.

Step 8. Modify the tester module as follows: Consider the **if** statement in *srtester.vhd* that compares the output of the D flip-flop with the test vector and writes *outfile.txt*. Replace this statement with an ASSERT statement as follows:

> **assert** Q = check(1) **and** Qbar = check(0)
> **report** "Test Vector Failed"
> **severity error**;

Step 9. Rerun the simulation. Any previous error messages should appear as an assertion violation accompanied by the error message being printed on the simulator command line rather than being directed to the file *outfile.txt*.

End Simulation Exercise 7.2

7.6   Chapter Summary

This chapter has presented the basic concepts used in reading and writing files. One common application of file I/O was developed: writing testbenches for testing VHDL simulation models. The exercises stressed basic binary and text I/O from files and the construction of functions for reading and writing other data types. The concepts covered in this chapter include the following:

- Files, file types, and file declarations.
- Basic operations for reading and writing text and binary files.
- Creation and use of procedures for reading and writing other data types to files.
- Construction and operation of testbenches.

Armed with an understanding of the basic issues of file I/O, we can proceed to ask the right questions to determine the I/O functions that are supported within any commercial simulator. This includes declaring and using any general I/O packages that may be publicly available, such as the TEXTIO package, or creating a new package as required to read and write application-specific data types.

Exercises

1. Write and test a model for a 16-bit shift register that is initialized to a value read from a file.

2. In models of complex components, such as memories in modern processors, accesses to memory must adhere to certain restrictions. For example, you generally cannot write to the area of memory that stores program instructions. Modify the memory model used in Simulation Exercise 7.1 to include an **assert** statement with a severity level of NOTE. Use this statement to produce a message whenever a particular memory location is written. Compile and test the model.

3. What are some of the advantages to using the idea of testbenches, rather than testing your model by providing stimuli to the entity input using the facilities in a simulator?

4. Consider a testbench for a combinational circuit such as the single-bit ALU described in Simulation Exercise 3.2. Write a VHDL model for this circuit. Develop a testbench to test the model. Verify the functionality of the ALU, using this testbench.

5. Modify the D flip-flop model shown in Figure 7.6 to include the use of an **assert** statement to check for conditions when both the Set and Reset inputs are asserted.

6. Write a testbench and test a model for an 8-bit counter. Use **bit** and **bit_vector** types. Therefore, you can use the procedures in the package TEXTIO to read test vectors from a file.

7. Consider the memory model of Figure 4.2. Modify this model to load memory from a file. Test this model by writing a testbench that accesses each memory location and compares the value with the original value in the file used to load memory.

8. Modify the testbench model of Figure 7.8 as follows:

   - The printed output should be formatted as follows:

     The test vector has failed. The model value of Q = 0 and the value of Qbar = 1
     The input vector is 11011

   - The printed output should be printed to a file and should appear as shown.

   - The problem is in printing the values of Q and Qbar, since they are of type std_logic. You can do so by using the case statement shown in Figure 7.3 in procedure write_v1d(). Since Q and Qbar are single-bit signals, the **for-loop** is not necessary.

CHAPTER 8

Simulation Mechanics

To this point, the text has focused on the VHDL language and the construction of simulation and synthesis models. Using these models, CAD tool environments support activities such as simulation, synthesis, and FPGA device programming. These activities are structured around a set of concepts defined within the language and a set of conventions defined by CAD tool vendors. We refer to this body of knowledge as the *simulation mechanics*. For example, for C or Pascal programs, we must be able to identify default libraries, program units that must be linked, and set the directory paths to be searched for missing program units.

This chapter covers the mechanics of organizing, building, and simulating VHDL models. Understanding the mechanics of these processes is useful in debugging and reasoning about model behavior, as well as quickly coming up to speed in being able to use VHDL-based CAD environments productively. The chapter provides an intuition about the practical aspects of VHDL environments and, hopefully, eases the transition into their proficient use and avoids some of the frustrations that often accompany navigation through a seemingly complex set of interrelated CAD tools. We first examine the practical aspects of simulation. The discussion strives to remain independent of specific CAD environments by focusing on issues common to all CAD environments. In this regard, the chapter seeks to convey a sense of being able to "ask the right questions" when attempting to navigate through a new CAD environment.

8.1 Terminology and Directory Structure

You will have noticed that much of the terminology used within the VHDL language has a hardware flavor—for example, signals, entities, and architecture. This way of thinking about VHDL designs continues through the design environment. The basic unit of

VHDL programming is a *design unit*, which is a component of a VHDL design and is one of the following:

- Entity
- Architecture
- Configuration
- Package Declaration
- Package Body

The entity, configuration declarations, and package declarations are primary design units—units that do not depend upon other design units. The remaining items are secondary design units. Design units are contained in *design files,* which contain the VHDL source for the individual design units. A single design file may contain the source for one or more design units. For example, imagine we were developing a VHDL model of an 8-bit adder. As newcomers to VHDL, our natural inclination is to place the **entity** and associated **architecture** code in the same file. This seems only natural. Alternatively, we could place the **entity** description and the **architecture** description in two separate files. When design entities are compiled or synthesized, CAD environments will create many intermediate files. The filenames are constructed from the names given to the design entities and not from the filenames. For example, suppose we created a file called *foo.vhd*, within which we place an **entity** named my_adder. When *foo.vhd* is compiled, intermediate files created will have names derived from my_adder. This is to be expected, since it is the design unit names and not the file names that link hierarchical models. (See Chapter 5.)

The compiled design units are placed in a *design library* in most computing environments, design libraries are typically implemented as directories. A design library is identified by a logical name, used to reference the library in a VHDL program using the **library** clause. The relationships of all of the components are illustrated in Figure 8.1. For the moment, think of the "analyzer" as the compiler. A better description will be provided in the next section. The figure shows the libraries with logical names WORK, STD, and IEEE. The library WORK is the current working library. If a user package is referenced in a user design unit, the compiled package must be placed in a designated library. The default library for the placement of the compiled package is the library WORK.

An arbitrary number of libraries can exist, and a given design can reference design units in one or more of these libraries as needed. Either the designer or the VHDL tools vendor may create a library. Two special libraries exist in all VHDL environments; the libraries STD and WORK. Recall that a library is implemented as a directory. The STD library contains the compiled descriptions of two packages: STANDARD and TEXTIO. The STANDARD package contains the definitions of the predefined types and functions of the language. For example, the definitions of **integer**, **real**, and **bit_vector** types and functions for operations on these types are defined in the package STANDARD. The TEXTIO package contains the predefined types, functions, and procedures for reading and writing from files using text-based I/O. The library

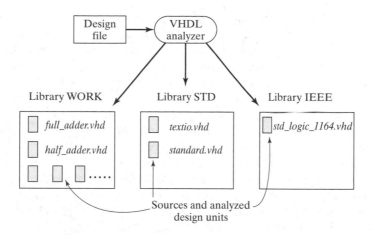

FIGURE 8.1   Placement and organization of design units and design libraries

WORK corresponds to your working directory. Analyzed design files are placed in the working directory. The logical names STD and WORK are defined by the system implementation. This means that if you wish to use packages that are available in these libraries, you can do so via the **use** clause. You do not need to declare these libraries with the **library** clause, as we do with the library IEEE. Typically, each design tool vendor will provide initialization files that assign directory path names to STD and WORK when you first invoke the VHDL environment. If you wish to use other libraries in your VHDL programs, you must decide on a logical name to reference each one. This logical name then must be set up in the host environment to point to the physical directory that will contain the packages that constitute the library. Check your simulator documentation for details.

8.2   Simulation Steps

Just as we have compilation, linking, and loading of C or Java programs, the major concepts in mechanics for simulation can be identified as *analysis, elaboration, initialization,* and *simulation*.

8.2.1   Analyzing VHDL Programs

Analyzing VHDL programs is synonymous with compiling VHDL programs. Indeed, the two terms are often used interchangeably. In some CAD tool environments, you will find menu items labeled Analyze, while others may use the term Compile for the same purpose. Consider the conventional process of compiling and executing a Pascal program.

We start with a text file containing the program. This program may reference functions or procedures found in other program modules, such as function libraries created by the user, or system libraries that are a part of the programming environment. For example, procedures to read and write files, as well as many mathematical functions, are provided by the system libraries, while the user may have created a library of functions for manipulating data such as strings or image data. Common programming environments provide rules for referencing these libraries and for compiling and linking independently compiled program modules to create a single executable program image. Similar conventions exist for compiling and linking distinct VHDL modules into a single image used for simulation. The structure of the programming units and the concepts that govern their compilation and linking are similar to those governing conventional programming languages. Design files containing design units are *analyzed* to produce a form that a simulator uses. The *VHDL analyzer* performs the customary syntactic checks and compilation to a form executable by a VHDL simulator. This process is analogous to the process of compilation of conventional programs such as Pascal.

With respect to analyzing VHDL programs, we must be concerned with the order in which design units are analyzed, because of dependencies between them. Throughout the examples in this text, entities and architectures were maintained in the same file for convenience. This is not strictly necessary. Design units can be in independent files and analyzed separately. Now, if we make a change to one file, what then is the order in which a large set of design entities must be analyzed? What dependencies between design units must we be aware of?

We can generate an intuition about compilation order by remembering that VHDL is a hardware description language. Consider the architecture of the structural model of a full adder reproduced from Figure 5.2 and illustrated in Figure 8.2. In order to build the circuit shown, we must first have "built," so to speak, the half-adder circuits and the two-input OR gate. Analogously, we must first have analyzed these design entities before we can analyze the architecture named structural. In general, for hierarchically structured models, we must remember to analyze hierarchical descriptions in a bottom-up fashion, starting from models at the lowest level in the hierarchy and proceeding to higher levels.

When we make changes to a design, we must consider dependencies between design units in determining which ones must be reanalyzed. Again, let us pursue the hardware analogy. Suppose we describe a board-level design by the interconnection of the component chips and the chip interfaces. If we now decide to replace one of the chips with a new chip having a different interface, then we must take a closer look at the design and reanalyze it to make sure it is still functionally and electrically correct. On the other hand, if we replace a chip with a newer, cheaper version, but one that maintains the same interface, we really do not have to perform any analysis, and the design should remain correct. Similar relationships carry over to the relationships among VHDL design units. Consider the structural model in Figure 8.2. This model depends on the availability of an entity named half_adder. If that entity description is changed (as in changing the chip interface), then the full_adder model that *depends* on it must be reanalyzed. Thus, if any entity description is changed, then *all* architectures that

```
architecture structural of full_adder is
component half_adder is
port (a, b : in std_logic;
      sum, carry : out std_logic);
end component half_adder;

component or_2 is
 port (a, b : in std_logic;
       c : out std_logic);
end component or_2;

signal s1, s2, s3 : std_logic;

begin
H1: half_adder port map (a => In1, b => In2, sum => s1, carry=> s3);
H2: half_adder port map (a => s1, b => c_in, sum => sum, carry => s2);
O1: or_2 port map (a => s2, b => s3, c => c_out);

end architecture structural;
```

FIGURE 8.2   Structural model of a full adder

depend on that entity description must be reanalyzed. For similar reasons, if a package declaration is changed, then any design unit (entity, architecture) that depends on that package must be reanalyzed. This chain of dependencies may pass through multiple design units.

However, suppose all of the entity descriptions and architecture descriptions are in separate files. What if we now change the architecture and not the entity description? Continuing with the natural hardware analogy, consider the architecture in Figure 8.2. Suppose we wish to change this architecture to use some new, detailed two-input OR gate component model named fast_OR2. The half-adder model is not affected, nor is the entity description of the full adder, which happens to be in a separate file. We simply need to (i) declare the component fast_OR2, (ii) change the component instantiation statement O1 to reflect the use of this new component, and (iii) reanalyze the architecture structural. Alternatively, suppose that the architecture of the half adder has changed, but the entity description of the half adder remains the same. In this case, we simply reanalyze the architecture of the half adder. We do not need to reanalyze the full-adder model! From our hardware analogy, it is as if we had wired up the full-adder circuit and simply swapped out the half-adder chip for another one. As long as the entity description of the half adder has not changed, the changes in the architecture of the half adder are not visible to the full adder at this time. When the full-adder model is loaded into a simulator, the later architecture for the half adder will be used.

At this point in our VHDL modeling experience, we may be inclined to organize our models with entity and architecture descriptions in the same physical file. Therefore, even if we change only the architecture description when we reanalyze the file, the entity description is also reanalyzed and thus appears to have changed! All models that use this entity now have to be reanalyzed! This is because the environment determines whether a design unit has changed by looking at the time stamp of the files created by the analyzer. If we reanalyze an entity description, even if we have not changed the interface, the creation time of the analyzed files will have changed. The VHDL environment must operate on the assumption that the entity description has changed. We now cannot simulate any model that uses this entity without reanalyzing it.

In sum, analyze design units in a bottom-up fashion. When you change entities or package declarations, you must reanalyze all design units that depend on those units. If you change the architecture or package bodies, then you need to reanalyze only those units. If just the architectures are to be analyzed, ensure that they are in separate files from the associated entity description to avoid unnecessarily propagating changes throughout the design hierarchy.

8.2.2   Elaboration of VHDL Programs

We have seen that structural models are a means for managing large designs. Before we can simulate a design, we must first flatten it into a description of the system that can be simulated. We know that a concurrent signal assignment statement is a process. Now we see that flattening a design essentially produces a large number of processes that communicate via signals, referred to as *nets*. It is easy to think of a combinational circuit as a set of nets. Consider the gate level circuit shown in Figure 8.3. The signals s1, s2, s3, s4, s5, s6, and z correspond to nets. A *netlist* is a data structure that describes all of the components connected to each net. Many industry standard formats exist for describing a netlist so that designs may be transferred easily between various design tools. This process of flattening a hierarchical description of a design is done during the phase of *elaboration* of the VHDL model. The elaboration produces a netlist of processes. Other functions are also carried out during elaboration, and the overall process is composed of the following steps:

1. *Elaboration of the Design Hierarchy:* This step includes flattening the hierarchy. In doing so, components must be associated with the architecture that is to be used to describe their behavior. This may be specified through a configuration or

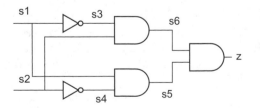

FIGURE 8.3   Netlists in a circuit

through the default choice of the architecture for an entity. Flattening the hierarchy produces a netlist of processes, each describing the behavior of each component at the lowest level of the hierarchy.

2. *Elaboration of the Declarations*: Recall that generic parameters must be constants and their values must be known prior to simulation. These values are determined and checked in this step. The declarations are also checked for type consistency and for whether the initialization of signals and variables is in accordance with the relevant rules of the language.

3. *Storage Allocation*: Storage is allocated for variables, signal drivers, constants, and other program objects.

4. *Initialization*: All signals and variables are initialized in this step, either to user-specified values, such as those provided in declarations, or to default values.

8.2.3 Initialization of VHDL Programs

Prior to simulation, two important actions take place. First, all signals (nets) are initialized to their default or explicitly initialized values. Second, all processes are executed until they are explicitly suspended by the use of wait statements or implicitly suspended by the use of a sensitivity list. This execution may also produce values for signals. The simulation time is set to 0 ns, and the model is ready to be begin simulation.

8.2.4 Simulation of VHDL Programs

Simulation proceeds as a discrete event simulation of the design. This is achieved by evaluating the values of all signals. If an event has occurred on any signal (i.e., net), all of the processes affected by that signal are executed, possibly generating future events on other nets. Conceptually, the events are managed as an event list organized according to the time stamp of the event as described in Chapter 2.

Let us consider what actually happens when a structural model, such as the one shown in Figure 8.2, is simulated. We might first create a file with this model. Let us call this file *structural.vhd*. The models for the half adder and the two-input OR gate may be created in two other files, say *ha.vhd* and *OR2.vhd*. We now know that the order of compilation is important. The latter two files are analyzed first. Independently of the file names, the compiled design units will be identified by their entity and architecture labels.

The process of simulation utilizes a number of concepts that are realized in various simulators in different ways. Some of the common steps that you can expect to encounter are discussed next.

Initialization The simulator environment must maintain information about various design units involved in the simulation, such as the location of libraries. We know that libraries are logical names for directories. Most simulators need access to information about the libraries that you plan to use (such as IEEE), the location of your working design library (WORK), and the location of the library containing the standard packages expected with the VHDL distribution (STD). Usually the simulation environment creates and maintains this information in initialization file. On some simulators, it may be necessary to set them explicitly by editing the initialization files created at installation time. If the VHDL analyzer returns errors relating to the absence of key libraries, it is most likely a result of the failure to define the physical location of the logical libraries.

Loading the Model   For the example in Figure 8.2, the compilation will produce a design unit named full_adder. Simulators will provide facilities for loading a model. When the design unit full_adder is loaded into a simulator, the working directory (WORK), IEEE, STD, and any other libraries that you may have declared will be searched for any compiled design units with the labels half_adder and or_2. The order in which these libraries are searched is important, since you do not want to inadvertently use a design entity of the same name in another library. This order of search is simulator specific, although it is not uncommon for the order to be based on that in which they are listed in the initialization files. Usually, you would want the library WORK to be the first in the search order.

In this example, since no configuration is explicitly provided, the simulator will look for a design unit with the same name as the component and an architecture associated with *that* component. In our example, the system would look for an entity named half_adder. Since no other information is provided, the names, types, and modes of all of the signals provided in the component declaration must exactly match those in the entity declaration. Recall that if you use the **port map** and **configuration** statements, this will not be necessary.

Simulation Setup   Now that the model is loaded, we may want to generate test cases and provide a stimulus to the model to determine whether it is indeed operating correctly. Generally, most simulators have ways in which to specify a waveform on an input port of the entity being loaded. There will also be facilities for forcing signals to a certain value. Signal initialization, especially on input ports, is a necessary prelude to simulation. If we have constructed a model with a testbench format, then we may not need to do so. However, if we load a model such as the full adder into the simulator, we will need to provide a stimulus to each of the inputs and examine the output signals to determine whether the model is correct.

Example: Generating an Input Stimulus

Figure 8.4 shows an example of the stimulator dialog box from the Active-HDL simulator. A waveform is specified on the input signal c_in by specifying the following formula:

1 0, 0 10 ns, 1 40 ns -r 60 ns

The formula can be interpreted as follows: Each pair of numbers specifies a value and a time. The formula states that the signal has the value 1 at time 0 ns, the value 0 at time 10 ns, and the value 1 at time 40 ns. The pattern repeats 60 ns later. By denoting

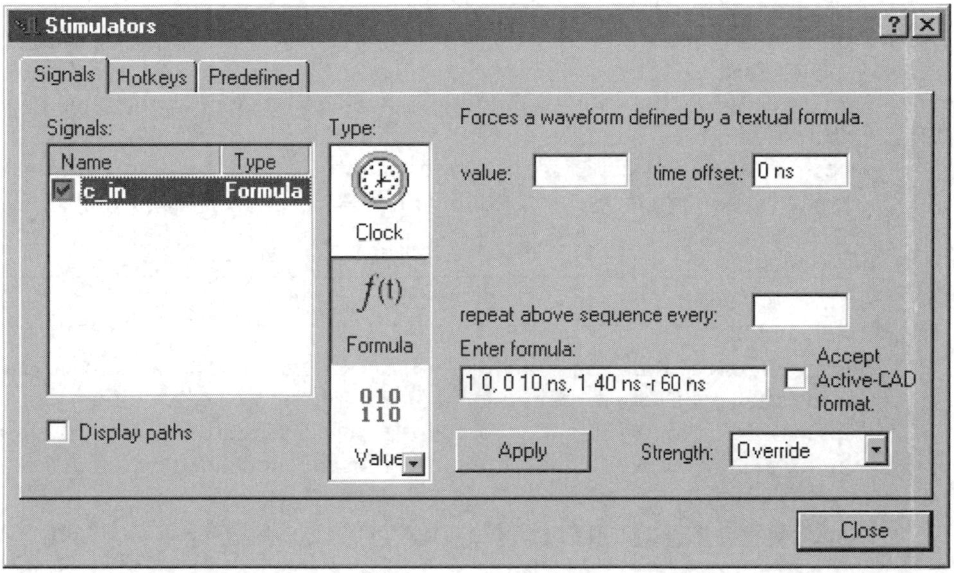

FIGURE 8.4   An example of creating stimulus on an input signal

this waveform as a clock pattern, as simulation time progresses, the waveform will be applied repeatedly until the user changes it. From the dialog box, we can see that other options are available, such as forcing a signal to a value and toggling the value of the signal with a "hot key" or defining clocks.

Example End: Generating an Input Stimulus

Another aspect of simulation is the notion of a simulator *step time*. Typically, you can advance simulation time in units called steps. For example, you might pick the step time as 10 ns. Running the simulation for 10 steps is analogous to running it for 100 ns. Simulators will provide facilities for setting the value of the step size. Step time can be important because the input stimulus to a model being tested often is applied for a period equal to the step time. For example, suppose you would like to create the following waveform on an input signal named reset: 0 1 0 0. That is, the signal reset should be driven to logic 0 for some duration, then logic 1 for some duration, and so on. For how long is reset driven to each value? The duration is generally equal to the simulator step time. If it is necessary to have a pulse of a specific width on reset, then the step time should be adjusted accordingly. More recently, simulators have been moving away from the notions of step time and deal with physical timing values.

Execution and Tracing   We are finally ready to execute the model and trace the values of signals. Typically, we can initiate execution by a run or step command. The former starts a simulation for a fixed period of time or a fixed number of simulator steps. The latter steps through a single simulation cycle. Typically, the values of all signals can be displayed in a trace window. *Remember, you cannot trace variables, only signals!* You can set break points and look at the value of the variables via simulator commands. However, variables do not exhibit time-dependent behavior in the same manner as signals and therefore cannot be traced in that fashion. Most simulators will also provide access to simulation statistics, such as the number of events executed. This can provide a useful insight into the behavior of the models, as well as aid in debugging them.

Example: Signal Traces

All simulators provide some facility for tracing signals so that we may actively monitor the internal behavior of the model. An example of a trace window from the Active-HDL simulator for a full-adder model is shown in Figure 8.5. By examining specific points on the trace, we can determine whether the model is functioning correctly.

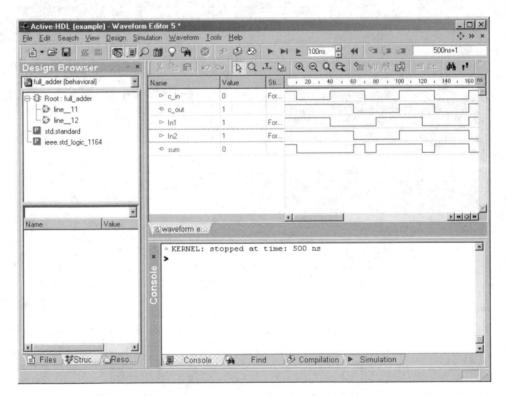

FIGURE 8.5   An example of a signal trace

Example End: Signal Traces

8.3   Chapter Summary

Traditionally, CAD tools have been large, complex aggregations of tools such as simulators, schematic capture tools, and layout editors. Although one can have a thorough knowledge of VHDL or VLSI design, and this knowledge may enable one to understand the concepts underlying the use of these tools, it also appears to be necessary to have some experience in the state of the practice to be able to use the tools successfully with minimal aggravation. This experience, often unrelated to design, is frequently referred to as "knowledge of the third kind." From a learning perspective, that situation is changing with the advent of low-cost, relatively easy-to-use point tools such as VHDL environments for PCs. Since these are focused on VHDL, we can make use of them without having to navigate through the maze of complex CAD environments. However, there are still conventions, concepts, and state-of-the-practice notions that are acquired through experience. It is difficult to provide concrete actions, since these are dependent on the VHDL toolset and its command set. However, we can introduce the common steps, concepts, and operations that will be encountered in the use of almost any simulator. Understanding how environments are structured and how these structures are related to language concepts is necessary for productive application to VHDL modeling. This chapter has attempted to provide an intuition about the practical aspects of VHDL environments and hopefully ease the transition into proficient use. Often, we can avoid being frustrated simply by knowing what questions to ask and what to expect. This chapter, coupled with detailed tutorials for specific VHDL environments provided in the appendices, is intended to bring the reader closer to that goal.

The concepts covered in the chapter include the following:

- Basic design units
  - entity
  - architecture
  - configuration
  - package declaration
  - package body
- Analyzing VHDL programs
  - order of analysis of design units
- Elaboration
  - elaboration of the design hierarchy
  - elaboration of the declarations
  - storage allocation
  - initialization

- Initialization of VHDL programs
- Simulation of a design

These concepts are embedded within VHDL environments, although user interfaces increasingly hide the majority of them from users.

CHAPTER 9 — Identifiers, Data Types, and Operators

The VHDL language provides a rich array of data types and operators, along with rules that govern their use. The goal of this chapter is to provide ready access to the syntax and semantics of commonly used data types and operators. The chapter is intended to serve more as a guide in writing your first VHDL programs, rather than as a comprehensive language reference source. The more advanced language features are not referenced here but can be found in a variety of excellent texts on the VHDL language or the VHDL Language Reference Manual (LRM). Familiarity with common programming language concepts and idioms is assumed.

9.1   Identifiers

Identifiers are used as variable, signal, or constant names, as well as names of design units, such as entities, architectures, and packages. A basic identifier is a sequence of characters that may be upper- or lowercase, the digits 0–9, or the underscore ("_") character. The VHDL language is not case sensitive. The first character must be a letter and the last character must not be "_". Therefore, Select, ALU_in, and Mem_data are valid identifiers, while 12Select, _start, and out_ are not valid identifiers. The valid identifiers listed are known as *basic identifiers*.

9.2   Data Objects

'87 vs. '93   In VHDL'87, there are three classes of objects: signals, variables, and constants. In VHDL'93, files are introduced as a fourth class of object. In VHDL'87, file types represent a subset of a variable object type. The type of signal, variable, or constant object determines the range of values that can be assigned to it.

A signal is an object that holds the current and possibly future values of the object. In keeping with our view of VHDL as a language used to describe hardware, signals are typically thought of as representing wires. They occur as inputs and outputs in port descriptions, as signals in structural descriptions, and as signals in architectures. The signal declarations take the following form:

signal *signal_name*: *signal_type*:= *initial_value*;

Examples include

signal status : std_logic:= '0';
signal data : std_logic_vector (31 **downto** 0);

Recall that signals differ from variables in that signals are scheduled to receive values at some point in time by the simulator. Variables are assigned during execution of the assignment statement. At any given time, multiple values may be scheduled at distinct points in the future for a signal. In contrast, a variable can be assigned only one value at any point in time. As a result, the implementation of signal objects must maintain a history of values and therefore requires more storage and exact higher execution time overhead than variables do.

Variables can be assigned a single value of a specific type. For example, an integer variable can be assigned a value in a range that is implementation dependent. A variable of type real can be assigned real numbers. Variables are essentially equivalent to their conventional programming language counterparts and are used for computations within procedures, functions, and processes. The declaration of a variable has the following form:

variable *variable_name*: *variable_type* := *initial_value*;

Examples include the following:

variable address: **bit_vector**(15 **downto** 0) := x"0000";
variable Found, Done: **boolean :=** FALSE;
variable index: **integer range** 0 **to** 10:=0;

The last declaration states that the variable index is an integer that is restricted to values between 0 and 10 and is initialized to the value 0.

Constants must be declared and initialized at the start of the simulation and cannot be changed during the course of the simulation. Constants can be any valid VHDL type. The declaration has the following form:

constant *constant_name*: *constant_type* := *initial_value*;

Examples include the following:

constant Gate_Delay: **Time**:= 2 ns;
constant Base_Address: **integer**:= 100;

The first declaration states that the constant is of type **time**. This is a type unique to hardware description languages. Just as an integer variable can be assigned only integer values, the values assigned to the constant Gate_Delay must be of type time, such as 5 ns, 10 ms, or 3 s. In the preceding example, Gate_delay is initialized to 10 ns.

9.3 Data Types

The type of a signal, variable, or constant object specifies the range of values it may take and the set of operations that can be performed on it. The VHDL language supports a standard set of type definitions, as well as permitting the definition of new types by the user.

9.3.1 The Standard Data Types

The standard type definitions are provided in the package STANDARD (see Appendix B.1) and include the types listed in Table 9.1. Note the definitions of **bit** and **bit_vector** types. From Chapter 2, we know that a simple 0/1 value system is not rich enough to describe the behavior of single-bit signals. This is why the community has moved towards standardization of a value system that multiple vendors can use. Such a standard is the IEEE 1164 value system, which is defined with the use of enumerated types.

9.3.2 Enumerated Types

Although the standard types are useful for constructing a wide variety of models, they fall short in many situations. We know that single-bit signals may be in states that cannot be represented by 0/1 values. For example, signal values may be unknown, or signals may left floating. The language does support the definition of new language types

TABLE 9.1   Standard data types provided within VHDL

Type	Range of values	Example declaration
integer	implementation defined	**signal** index: **integer**:= 0;
real	implementation defined	**variable** val: **real**:= 1.0;
boolean	(TRUE, FALSE)	**variable** test: **boolean**:=TRUE;
character	defined in package STANDARD	**variable** term: **character**:= '@';
bit	0, 1	**signal** In1: **bit**:= '0';
bit_vector	array with each element of type bit	**variable** PC: **bit_vector**(31 **downto** 0)
time	implementation defined	**variable** delay: **time**:= 25 **ns**;
string	array with each element of type character	**variable** name : **string**(1 **to** 10) := "model name";
natural	0 to the maximum integer value in the implementation	**variable** index: **natural**:= 0;
positive	1 to the maximum integer value in the implementation	**variable** index: **positive**:= 1;

by the programmer and the ability to provide functions for operating on data that are of this type. For example, consider the following definition of a single bit:

type std_ulogic **is** ('U' -- uninitialized
                        'X' -- forcing unknown
                        '0' -- forcing 0
                        '1' -- forcing 1
                        'Z' -- high impedance
                        'W' -- weak unknown
                        'L' -- weak 0
                        'H' -- weak 1
                        '-' -- don't care
                        );

Now, assume we declare a signal to be of this type.

    signal carry: std_ulogic : = 'U';

The signal carry can now be assigned any one of the values defined in the preceding statement. Note that operations such as AND, OR, and "+/-" must be redefined for this data type. You can provide the type definitions and the associated operator and logical function definitions in a package that is referenced by your model. The preceding type definition is a standard defined by the IEEE and the associated package is referred to as the IEEE Standard Logic 1164 package (see Appendix B.3). This package is popular for two reasons: It provides a type definition for signals that is more realistic for real circuits, and the use of the same value system makes it easier for designers to share models, increasing interoperability and consequently reducing model cost.

    The preceding type definition is referred to as an *enumerated type*. The definition explicitly enumerates all possible values that a variable or signal of this type can assume. Another example in which enumerated types come in handy is the following:

    type instr_opcode **is** ('add', 'sub', 'xor', 'nor', 'beq', 'lw', 'sw');

    An instruction set simulation of a processor may have a large case statement with the following test:

    case opcode **is**
    when beq =>

Each branch of the **case** statement may call a procedure to simulate the execution of that particular instruction. Such type declarations can be made and placed in a package or in the declarative region of the **process**. (For an example of the definition, declaration, and use of a type definition for memory in a simulation model of a simple processor, see the example code in Figure 4.2.)

9.3.3   Array Types

Arrays of bit-valued signals are common in digital systems. An array is a group of elements, all of the same type. For example, a word is an array of bits and memory is an array of words. A common practice is to define groups of interesting digital objects as a new type. For example,

type byte **is array** (7 **downto** 0) **of bit**;
type word **is array** (31 **downto** 0) **of bit**;
type memory **is array** (0 **to** 4095) **of** word;

Now that we have created these new types, we can declare variables, signals, and constants to be of those types:

signal program_counter : word;= x"00000000";
variable data_memory : memory;

The use of type definitions in this manner enables us to define elements that we use in designing digital systems. For example, the preceding declarations demonstrate how we could define new types for words, registers, and memories. These new types make writing VHDL models more intuitive, as well as easier to comprehend.

9.3.4   Physical Types

Physical types are motivated by the need to represent physical quantities such as time, voltage, or current. The values of a physical type are defined to be a measure such as seconds, volts, or amperes. The VHDL language provides one predefined physical type: **time**. The definition of the type **time** can be found in the package STANDARD. The definition from this package is reproduced here:

type time **is range** *<implementation dependent>*
units
fs;   -- femtoseconds
ps = 1000 fs;   -- picoseconds
ns = 1000 ps;   -- nanoseconds
us = 1000 ns;   -- microseconds
ms = 1000 us;   -- milliseconds
s = 1000 ms;   -- seconds
min = 60 s;   -- minutes
hr = 60 min;   -- hours
end units;

The first unit is referred to as the base unit and is the smallest unit of time. All of the other units can be defined in terms of any of the units defined earlier. For example, they all could have been defined in terms of femtoseconds. We can now see how it is possible to define other physical types, such as distance, power, or current. For example, we might define a type power as follows:

type power **is range** 1 to 1000000
units
uw;   -- base unit is microwatts
mw = 1000 uw;   -- milliwatts

```
w = 1000 mw;          -- watts
kw = 1000000 mw;      -- kilowatts
mgw = 1000 kw;        -- megawatts
end units;
```

Note how kilowatts are defined in terms of milliwatts rather than watts. Now we can declare signals or variables to be of this type and assign or compute values of that type:

```
variable chip_power: power:= 120 mw;
```

The declaration creates a variable of type power whose value will be in the units defined for variables of type power. When we are modeling physical systems, it is very useful to have the ability to define physical types and the units used to express their values. We might use such types to execute simulations that estimate the power dissipation over the course of execution of a component. This capability merits an example.

Example: Use of Physical Types

Consider the code shown in Figure 9.1. Let us assume that we are using the definition of the type power, as shown earlier in this section, and that this definition is placed in

```
library IEEE;
use IEEE.std_logic_1164.all;
use Work.my_pkg.all;

entity example_power is
port(clk : in std_logic;
outpower : out power);
end entity example_power;

architecture behavioral of example_power is

begin
process is
variable my_power: power:= 1 uw;

begin
wait until (rising_edge(clk));
--
-- miscellaneous modeling code here
--
my_power := my_power + 100 uw;
outpower <= my_power;
end process;
end architecture behavioral;
```

FIGURE 9.1 An example of the use of physical types

a package named my_pkg which is compiled into the local working directory. The code template shown in Figure 9.1 illustrates how one might use physical types. In this case, each time the circuit modeled by the code is executed, the power dissipated is increased. This value may be written to a file or, as shown, written to a port. Thus, higher level components in a hierarchy may use this computed value or simply drive a signal trace for viewing purposes.

Example End: Use of Physical Types

9.4   Operators

VHDL'93

☞

Operators are used in expressions involving signal, variable, or constant object types. The table that follows lists the sets of operators, as defined in the VHDL language [8]. Note that the shift operations are new in VHDL'93. The miscellaneous operators include **abs** for the computation of the absolute value, and ** for exponentiation. The latter can be applied to any integer or real signal, variable, or constant. Here is the table:

logical operators	**and**	**or**	**nand**	**nor**	**xor**	**xnor**
relational operators	**=**	**/=**	**<**	**<=**	**>**	**>=**
shift operators	**sll**	**srl**	**sla**	**sra**	**rol**	**ror**
addition operators	**+**	**–**	**&**			
unary operators	**+**	**–**				
multiplying operators	*	**/**	**mod**	**rem**		
miscellaneous operators	**	**abs**	**not**			

The classes of operators shown are listed in increasing order of precedence from top to bottom. Operators with higher precedence are applied to their operands first. All of the operators within the same class are of the same precedence and are applied to operands in textual order—left to right. Parentheses can be used to define explicitly the order of precedence. In general, the operands of these operators must be of the same type, whereas the type of the permissible operands may be limited. Tables 9.2 through 9.7 provide information from the VHDL Reference Manual [8] and define the permissible operand types.

TABLE 9.2   Operator–operand relationships

Operator	Operand Type	Result Type
=	Any type	Boolean
/=	Any type	Boolean
<, >, <=, >=	Scalar or discrete array types	Boolean

Table 9.3 defines the shift operators. These operators are new in VHDL'93 and are particularly useful in describing operations in models of computer architecture components. Table 9.4 gives some examples of the application of these shift operators.

Table 9.5 describes the addition, subtraction, and concatenation operators. The first two are self explanatory.

The concatenation operator composes operands. For example, we might have:

result(31 **downto** 0) <= '0000' **&** jump (27 **downto** 0);

The upper four bits of result will be cleared, and the remaining 28 bits will be set to the value of the 28 least significant bits of jump.

Table 9.6 describes the unary operators.

Finally, Table 9.7 illustrates some remaining numerical operators.

'87 vs. '93

TABLE 9.3   Shift operators

Operator	Operation	Left operand type	Right operand type	Result type
sll	Logical left shift	Any one-dimensional array type whose element type is **bit** or **Boolean**	**integer**	left operand type
srl	Logical right shift	"	"	"
sla	Arithmetic left shift	"	"	"
sra	Arithmetic right shift	"	"	"
rol	Rotate left logical	"	"	"
ror	Rotate right logical	"	"	"

TABLE 9.4   Examples of the application of the shift operators

Example expression	Result operand value
value <= "10010011" sll 2	"01001100"
value <= "10010011" sra 2	"11100100"
value <= "10010011" ror-3 (i.e., rotate left)	"10011100"
value <= "10010011" srl-2	"01001100"

TABLE 9.5   Addition and subtraction operators

Operator	Operation	Left operand type	Right operand type	Result type
+ or -	Addition/subtraction	Numeric type	Same type	Same type
&	Concatenation	Array or element type	Array or element type	Same array type

TABLE 9.6   Unary operators

Operator	Operation	Operand type	Result type
+/-	Identity/negation	Numeric type	Same type

TABLE 9.7   Numerical operators

Operator	Operation	Left operand type	Right operand type	Result type
* or /	Multiplication or division	Integer or floating point type	Same type	Same type
mod or rem	Modulus/remainder	Integer type	Same type	Same type

9.5   Chapter Summary

This chapter has provided a brief overview of the common objects, types, and operators in the VHDL language. Its approach has focused on the unique aspects of the VHDL language throughout most of the text, while assuming that language features such as types, identifiers, and objects are familiar concepts to the reader and that a handy reference is all that is needed as a prelude to writing useful models. The chapter is intended to fill that role and will be only as useful as the previous chapters have been successful in providing an intuitive way of thinking about, constructing, and using VHDL models. Familiarity at this level can lead to the next level in using the more powerful (and complex) features of the language.

References

1. Active-HDL Reference, http://www.aldec.com/ActiveHDL/.

2. J. Bhaskar, *A VHDL Primer*. Englewood Cliffs, NJ: Prentice Hall, 1995.

3. J. Bhaskar, *A VHDL Synthesis Primer*. Allentown, PA: Star Galaxy Publishers, 1996.

4. K. C. Chang, *Digital Systems Design with VHDL and Synthesis: An Integrated Approach*. IEEE Computer Society, 1999.

5. B. Cohen, *VHDL Coding Styles and Methodologies*. Boston: Kluwer Academic, 1995.

6. D. Gajski and R. H. Kuhn, "Guest Editors Introduction—New VLSI Tools," *IEEE Computer*, vol. 16, no. 2, 1983, pp. 14–17.

7. J. Hayes, *Introduction to Digital Logic*. Reading, MA: Addison-Wesley, 1993.

8. *IEEE Standard VHDL Language Reference Manual: ANSI/IEEE Std 1076–1993*. New York: IEEE, June 1994.

9. Handel H. Jones, "How to Slow the Design Cost Spiral," Electronics Design Chain, September 2002, www.designchain.com.

10. R. Lipsett, C. Schaefer, and C. Ussery, *VHDL: Hardware Description and Design*. Boston: Kluwer Academic, 1989.

11. V. K. Madisetti, "Rapid Digital System Prototyping: Current Practice and Future Challenges," *IEEE Design and Test*, Fall 1996, pp. 12–22.

12. V. K. Madisetti and T. W. Egolf, "Virtual Prototyping of Embedded Microcontroller-Based DSP Systems," *IEEE Micro*, 1995, pp. 9–21.

13. D. Patterson and J. Hennessey, *Computer Organization & Design: The Hardware/Software Interface*. San Francisco: Morgan Kaufmann, 1994.

14. D. Perry, *VHDL*. 2d ed. New York: McGraw-Hill, 1994.

15. M. Richards, "The Rapid Prototyping of Application—Specific Signal Processors Program," *Proceedings of the First Annual RASSP Conference*, Defense Advanced Research Projects Agency, 1994.

16. K. Skahill, *VHDL for Programmable Logic*. Reading, MA: Addison-Wesley, 1996.

17. D. J. Smith, *HDL Chip Design: A Practical Guide for Designing, Synthesizing, and Simulating ASICs and FPGAs Using VIIDL or Verilog*. Doone Publications, 1996.

18. D. E. Thomas, C. Y. Hitchcock III, T. J. Kowalski, J. V. Rajan, and R. A. Walker, "Automatic Data Path Synthesis," *IEEE Computer*, vol. 16, no. 12, December 1983, pp. 59–70.

19. R. Walker and D. E. Thomas, "A Model of Design Representation and Synthesis," *Proceedings of the 22nd ACM/IEEE Design Automation Conference*, 1985, pp. 453–459.

20. S. Yalamanchili, *Introductory VHDL: From Simulation to Synthesis*. Upper Saddle River, NJ: Prentice Hall, 2001.

APPENDIX A Active-HDL Tutorial

The Active-HDL™ environment is tightly integrated into the Xilinx Foundation toolset and represents, at the current time, the state of the art in functional VHDL simulators. By functional we refer to the fact that the models may not necessarily be synthesizable. In a typical design methodology, functional VHDL simulation is used early in the design process to verify the correctness of a model and to generate test vectors for subsequent testing of the synthesized design.

A.1   Using Active-HDL

This tutorial assumes that we are starting with the creation of a new design. The sequence of steps proceeds through the creation and simulation of a VHDL model. The tutorial is for Active-HDL 6.2 (http://www.aldec.com). Student versions of the simulator are available.

Step 1: Creating a Project

In most VHDL environments, the first step is to create a workspace and, within that workspace, a project that will serve as the repository for the design files, intermediate files, and information concerning the options for the CAD tool environment. The Active-HDL environment maintains an analogous concept for managing a design.

When you first start Active-HDL, you will be prompted to open an existing workspace or to create a new one. Create a new workspace and name this workspace StartersGuide. You will now be prompted with a dialog box that will ask about adding existing resource files or creating an empty design. Select the option to create an empty design. Accept the defaults in the next dialog box and provide the design name

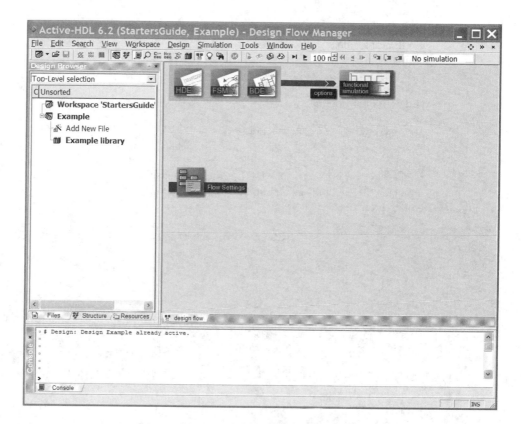

FIGURE A.1   Active-HDL environment on start-up

in the third dialog box. Call this design **Example**. The Active-HDL window should appear as shown in Figure A.1. (A default message may appear in the upper part of the window, rather than an empty canvas.)

Step 2: Creating a VHDL Model

The workspace for the design has been created in the form of directories, and all other environment variables have been initialized. We can now add files to this design. From the menu, select **File->New->Text Document**. This will open the Active-HDL text editor. Type in the full-adder model shown in Figure A.2. Save the model as the file *full.vhd*. The default directory in which the file is saved should be **Example/src**, where **Example** is the design directory in the workspace **StartersGuide**. There is a distinction between files and directories. You could directly create a VHDL text file in the directory corresponding to the design. However, the VHDL model is not part of the design until you explicitly add it to the design, by selecting **Design->Add Files to Design** from the menu and adding the file *full.vhd*.

This file should now appear in the **Design Browser** portion of the Active-HDL window with "?" next to the file name, as shown in Figure A.3.

```vhdl
library IEEE;
use IEEE.std_logic_1164.all;
entity full_adder is
port (In1, In2, c_in : in std_logic;
        sum, c_out : out std_logic);
end entity full_adder;

architecture behavioral of full_adder is

begin
sum <= In1 xor In2 xor c_in; VHDL is not case sensitive
c_out <= (In1 and In2) or (in1 and c_in) or (in2 and c_in);
end architecture behavioral;
```

FIGURE A.2   VHDL model of a full adder

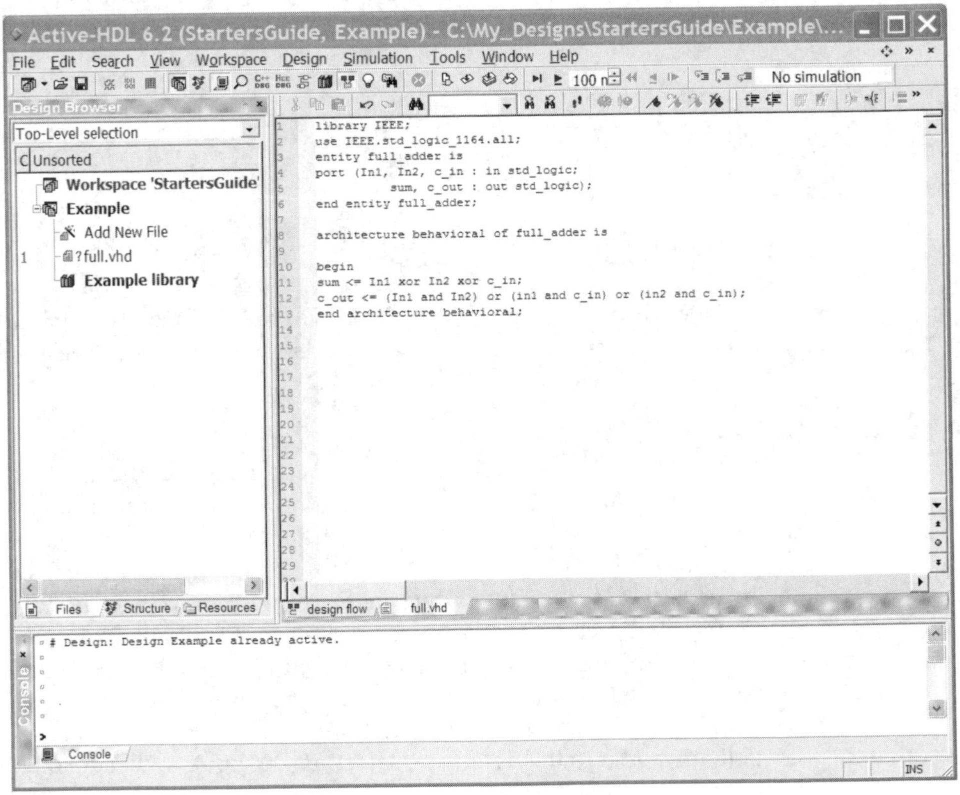

FIGURE A.3   Active-HDL environment with source file ready for compilation

Alternatively, we could have created a model by selecting File->New->VHDL Source from the file menu. This option would have started the source file wizard, which provides a convenient graphical interface to specify the names of the entity, architecture, and files, as well as the name and mode of all interface signals. The wizard can then create templates of the entity and architectures, and we need only fill in the code bodies manually. If we use this approach, the file can automatically be added to the design.

Step 3: Compiling a Design

After we have edited a text document to create the design, the file name shows up in the Design Browser with a "?" next to the filename to denote the fact that the file should be compiled. Each time the file is edited, the "?" will appear next to the filename in the design browser. We can compile the design by selecting Design->Compile from the menu. Any error messages will show up on the console window. After compilation, the Active-HDL window should appear as shown in Figure A.4. If there are compilation errors, you can use the text editor to correct the model, save the design, and then recompile the design. Complete this loop until the design compiles without errors.

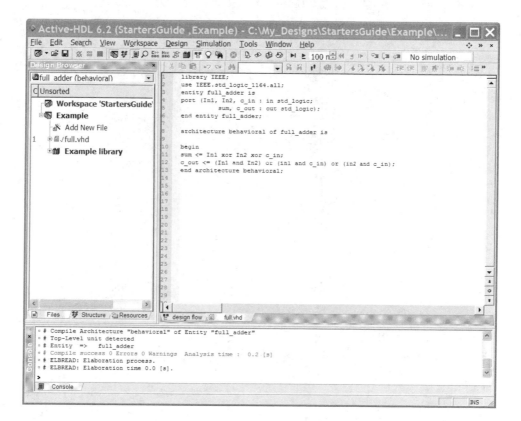

FIGURE A.4   The Active-HDL window after the design has been successfully compiled

It would be useful to create an error, such as a typographical error, and go through the steps of correcting and recompiling a design.

Step 4: Simulating a Design

We are now ready to simulate a design. In general, we may have many files in our design and may wish to simulate only one of then. We should first select the top level design that we wish to simulate. We can do so in the Design Browser part of the Active-HDL window by selecting the module full_adder(behavioral) from the pull-down menu. Note the tabs at the bottom of the Design Browser - Files, Structure, and Resources. Select each tab and you will find a different view of the design. For example, if we wish to make some changes in the source file and recompile the design, we would clearly wish to have a Files view. In particular, select the Structure tab. Click on the "+" next to the full_adder(behavioral) text to expand the selection. After we do so, the Design Browser window should appear as shown in Figure A.5.

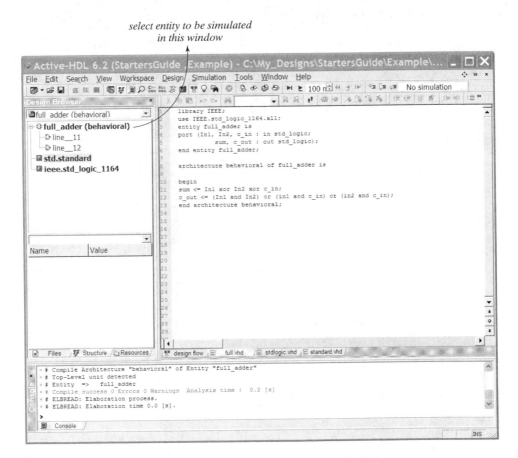

FIGURE A.5   Selecting a design entity for simulation

Simulation can verify the functional properties of a design. How can we verify that a circuit does what we think it does (or want it to do)? Design verification and validation is an increasingly large percentage of the design cycle for digital systems. A common technique is to provide input values and compare the output values with known correct output values. Generating such input vectors and associated correct output vectors is a complex task with any but the simplest of systems. Like most simulators, Active-HDL provides the ability to test simulation models in this fashion and in fact supports the automatic generation of testbench templates. (Check the vendor documentation for this capability, which is omitted in this quickstart tutorial.)

Since the circuit is simulated by stimulating the input signals with a known sequence of values and examining the outputs, we would like to be able to visualize the waveforms on the input and output signals. The basic activities in using most any simulator at this point, and the corresponding Active-HDL commands, are the following:

- *Select the signals to be traced*: From the menu, select File->New->Waveform. This will open a trace window in the Active-HDL window. You will notice that the trace window is empty. We need to specify the signals we wish to trace. We can do so by selecting Waveform->Add Signals from the menu. This will produce the dialog box shown in Figure A.6. Select each signal and click Add. For this exercise, add all of the signals to the trace. Once you have added all of the signals, the Active-HDL window will appear as in Figure A.7.

- *Generate the stimulus for the input signals*: The next step is to generate a stimulus on each of the input signals. Select signal In1 in the trace window, and then select Waveform->Stimulators from the menu. This will open the stimulator window

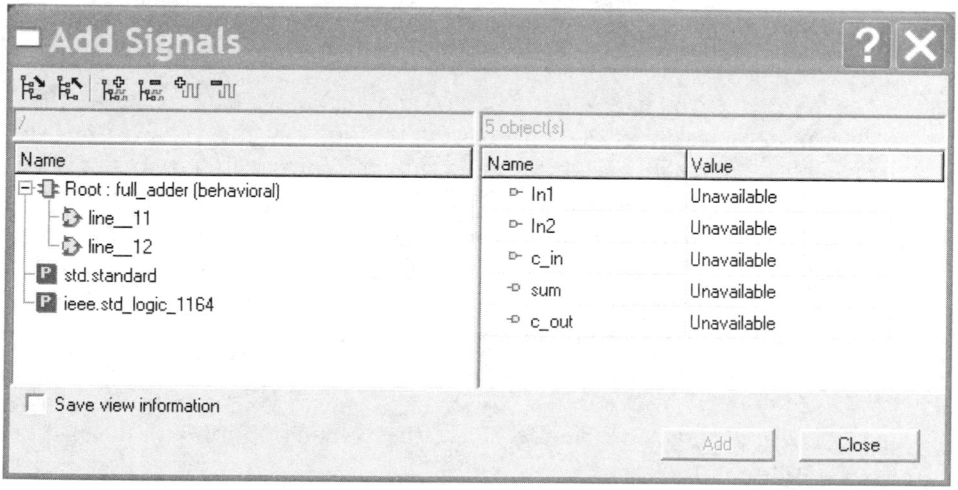

FIGURE A.6   Dialog box to add signals to the trace

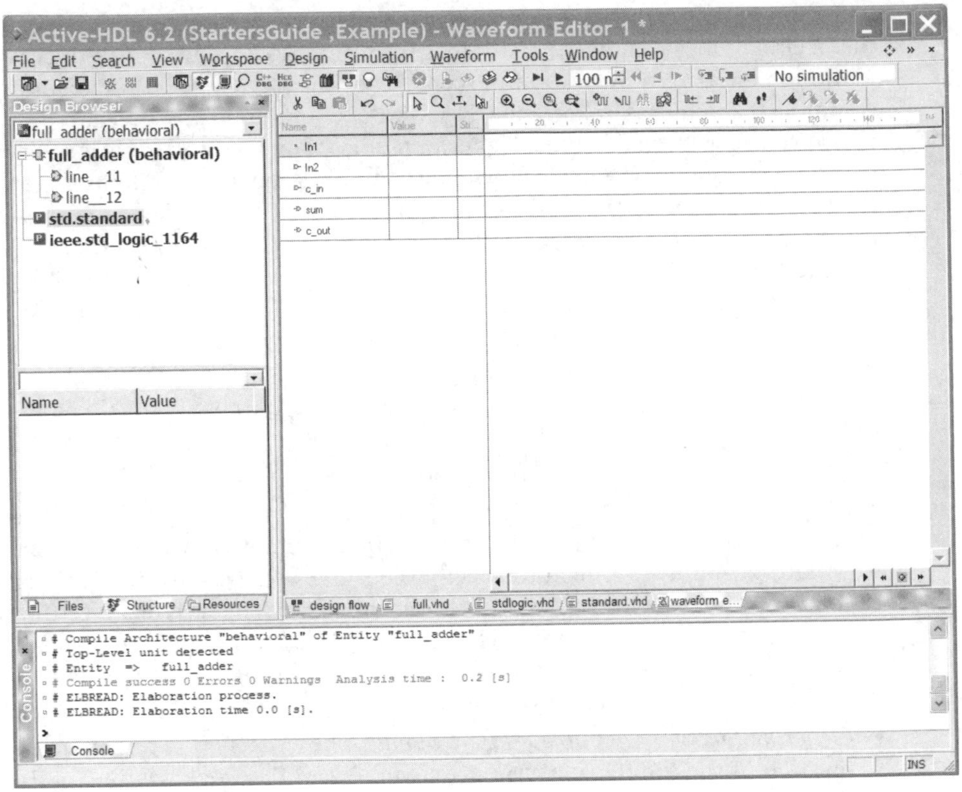

FIGURE A.7 Active-HDL with trace window ready for simulation

shown in Figure A.8. There are several ways in which to stimulate the values of an input signal.

- *Hot Key*: You can select a key on the keyboard that can toggle the value of an input signal. Whenever you press the hot key, the value of the signal toggles between 0 and 1. Thus, you can simulate for, say, 100 ns, toggle the value of an input signal, and then simulate for another 100 ns.

- *Clock Signals*: What if we chose to generate an arbitrary clock signal? How can we specify the pulse widths and the pulse separations? We can do so by clicking the clock button at the top of the list of options in the stimulator window. We can edit the clock information for any desired frequency. We can also select any duty cycle by clicking and dragging the rising edge of the clock signal shown in the window.

- *Formula*: In this example, select the formula option and fill in the formula as follows:

$$1\ 0, 0\ 10\ \text{ns}, 1\ 40\ \text{ns}\ \text{-r}\ 60\ \text{ns}$$

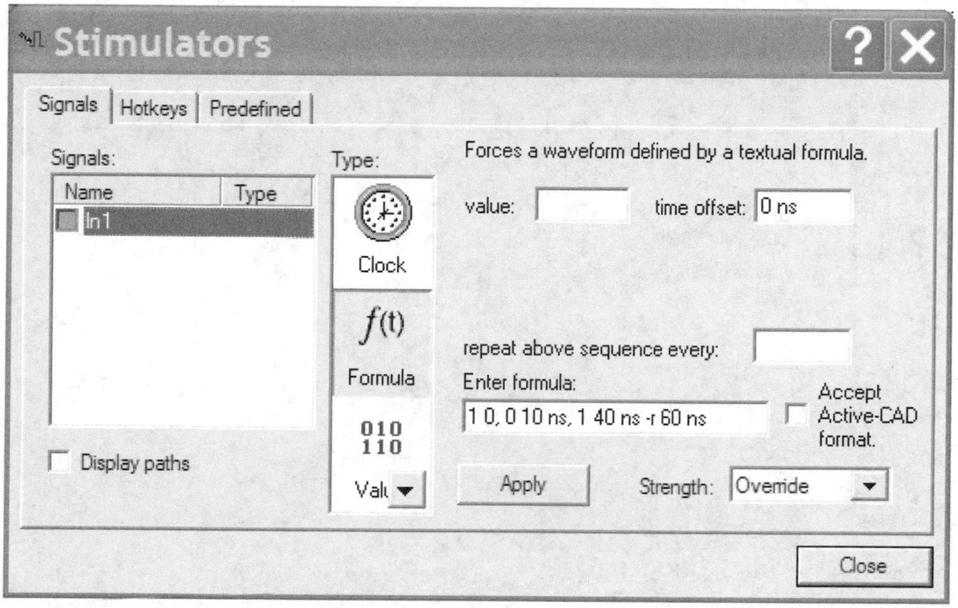

FIGURE A.8   The Stimulator dialog box

The formula can be interpreted as follows: Each pair of numbers specifies a value and a time. The preceding formula states that the signal has the value 1 at time 0 ns, the value 0 at time 10 ns, and the value 1 at time 40 ns. The pattern repeats 60 ns later.

- For signal In2, select the clock option and select a waveform with a 100-ns period and a duty cycle of 50%. (This should be the default.)
- For signal c_in, select the value option (shown as a sequence of binary numbers) and force the value of c_in to be 0.
- Note that there are several other options for specifying the values of a signal, including constant values, as well as some predefined clock waveforms.

Now we are ready to start the simulation. Under the Simulation menu item, we will find options to Initialize the simulation, End Simulation, and Run Until a specified period of time. Run the simulation for 500 ns. Now examine the trace window. It should appear as shown in Figure A.9. By examining the values of the signals in the trace window, we can determine whether the model is functioning correctly.

A.2   Miscellaneous Features

There are many useful features of the Active-HDL environment that go beyond what we find useful in a quickstart tutorial and that can be found in the online help or vendor documentation. Two of these features are particularly noteworthy.

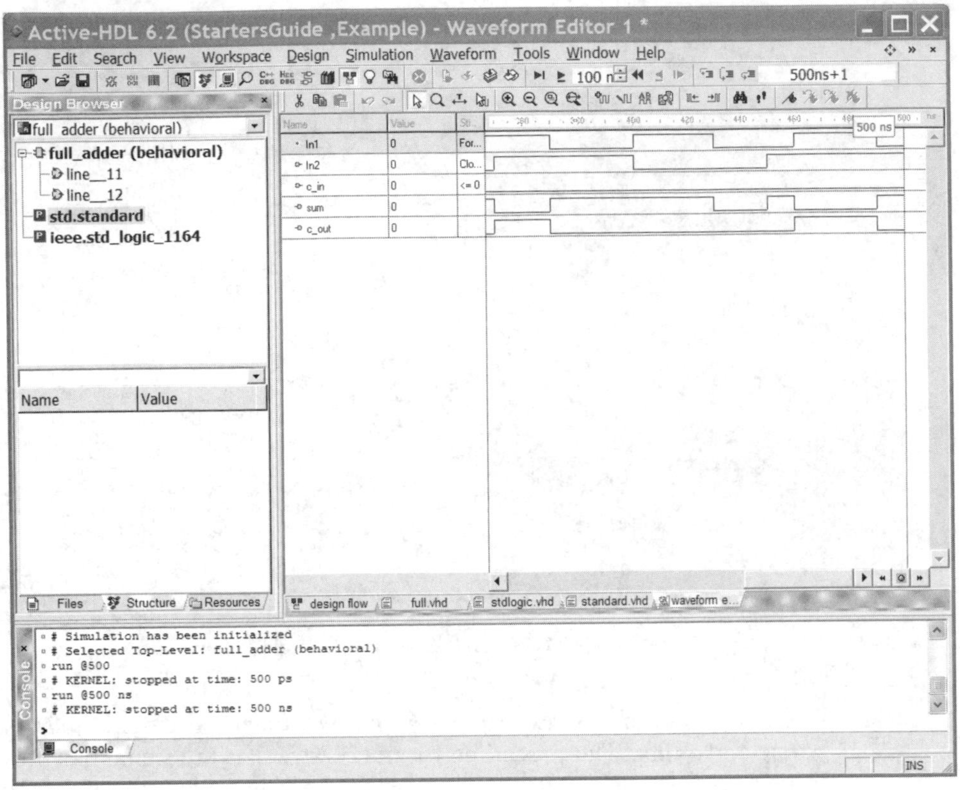

FIGURE A.9   Trace window after simulation

The first is the language assistant, which you can invoke by selecting Tools->Language Assistant from the menu. This will provide a dialog box with a menu of language constructs. For example, Figure A.10 shows the contents of the dialog box after selecting a for-loop from the list of Language Templates in the dialog box. We can cut and paste from the window into our text editor. A wide variety of templates is available for simulation. This Language Assistant is a handy reference and is enormously helpful in developing VHDL models.

A second particularly noteworthy feature is the automatic generation of a testbench for the design. The selection of Tools->Generate Test Bench from the menu will invoke the Test Bench Generator Wizard, which will guide you through the creation of a testbench template, including the use of a file that contains the test vectors.

A.3   Chapter Summary

This chapter has focused on a quickstart tutorial for Active-HDL version 6.2. I have presented the minimal set of activities or steps necessary to exercise the environment to simulate a single design. Early in digital logic and computer architecture courses,

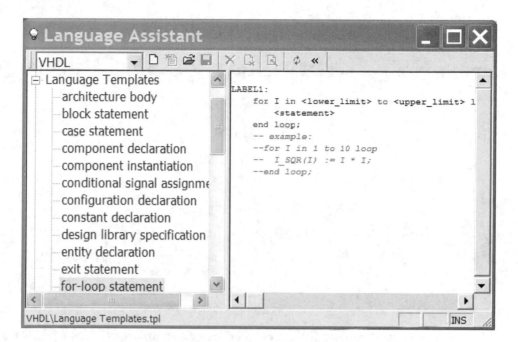

FIGURE A.10 Language Assistant

students are more likely to be constructing single monolithic models of relatively simple components. The tutorial is intended to help them get to the simulation of these monolithic models quickly and as painlessly as possible, so that they can concentrate on the digital logic and computer architecture concepts they are learning. It is hoped that the resulting familiarity will breed confidence to venture into the depths of these toolsets as necessary for the students to avail themselves of the powerful capabilities these environments have to offer. The steps covered here include the following:

- Creating a design
- Adding VHDL files to a design
- Compiling VHDL files
- Simulating the design

  - generating a stimulus on input signals
  - creating a trace of the input and output signals

- Use of the Language Assistant

The environment provides for many more advanced design activities described in the vendor documentation.

Standard VHDL Packages

This appendix contains listings of interfaces to standard packages available with the VHDL distributions. The STANDARD and TEXTIO packages are provided as part of the implementation of the VHDL environment. The package std_logic_1164 is an implementation of the IEEE 1164 value system and is generally provided by all vendors to support the generation of portable VHDL models. Several other packages are also typically made available by vendors. The reader is encouraged to browse through the package headers at your installation and study the contents.

B.1 Package STANDARD

All vendors distribute the package STANDARD. This package provides the definitions of the predefined types and functions for the language. The package header contents shown here are from the 1993 IEEE Standard VHDL Language Reference Manual [8]. The implementation of this package will be consistent across all vendors. The listing of the package header shown next omits the definition of all of the operators for each type.

```
--------------------------------------------------------------------------------
-- ANSI/IEEE Std 1076–1993*
-- IEEE Standard VHDL Language Reference Manual
-- Copyright ©1993 by the Institute of Electrical and Electronics Engineers, Inc.
-- The IEEE disclaims any responsibility or liability resulting from the placement
-- and use in the described manner. Information is reprinted with the permission
```

```
-- of the IEEE
--
----------------------------------------------------------------------------
--
--Predefined enumeration types
--
package STANDARD is
  type BOOLEAN is (FALSE, TRUE);
  type BIT is ('0', '1');
  type CHARACTER is (
    NUL, SOH, STX, ETX, EOT, ENQ, ACK, BEL,
    BS, HT, LF, VT, FF, CR, SO, SI,
    DLE, DC1, DC2, DC3, DC4, NAK, SYN, ETB,
    CAN, EM, SUB, ESC, FSP, GSP, RSP, USP,
    ' ', '!', '"', '#', '$', '%', '&', ''',
    '(', ')', '*', '+', ',', '-', '.', '/',
    '0', '1', '2', '3', '4', '5', '6', '7',
    '8', '9', ':', ';', '<', '=', '>', '?',
    '@', 'A', 'B', 'C', 'D', 'E', 'F', 'G',
    'H', 'I', 'J', 'K', 'L', 'M', 'N', 'O',
    'P', 'Q', 'R', 'S', 'T', 'U', 'V', 'W',
    'X', 'Y', 'Z', '[', '\', ']', '^', '_',
    ''', 'a', 'b', 'c', 'd', 'e', 'f', 'g',
    'h', 'i', 'j', 'k', 'l', 'm', 'n', 'o',
    'p', 'q', 'r', 's', 't', 'u', 'v', 'w',
    'x', 'y', 'z', '{', '|', '}', '~', DEL);
--
-- there are a host of other characters here including some special characters which
-- are omitted from this presentation
--
  type SEVERITY_LEVEL is (NOTE, WARNING, ERROR, FAILURE);
  type universal_integer is range implementation_defined;
  type universal_real is range implementation_defined;
--
-- in implementations of this package the statement below would define
-- numeric values in the range field
--
  type INTEGER is range implementation_defined;
  type REAL is range implementation_defined;
  type TIME is range implementation_defined
```

```vhdl
  units
    fs;                          -- femtosecond
    ps = 1000 fs;                -- picosecond
    ns = 1000 ps;                -- nanosecond
    us = 1000 ns;                -- microsecond
    ms = 1000 us;                -- millisecond
    sec = 1000 ms;               -- second
    min =   60 sec;              -- minute
    hr =   60 min;               -- hour
  end units;
subtype DELAY_LENGTH is TIME range 0 fs to TIME'HIGH;
-- function that returns the current simulation time:
impure function NOW return DELAY_LENGTH;
  -- predefined numeric subtypes:
  subtype NATURAL is INTEGER range 0 to INTEGER'HIGH;
  subtype POSITIVE is INTEGER range 1 to INTEGER'HIGH;
  -- predefined array types:
  type STRING is array ( POSITIVE range <> ) of CHARACTER;
  type BIT_VECTOR is array ( NATURAL range <> ) of BIT;
--
--predefined types for opening files
--
type FILE_OPEN_KIND is (READ_MODE, WRITE_MODE, APPEND_MODE);
type FILE_OPEN_STATUS is (OPEN_OK, STATUS_ERROR, NAME_ERROR,
        MODE_ERROR);

attribute FOREIGN: STRING;

end STANDARD;
```

B.2   Package TEXTIO

The TEXTIO package is distributed by all vendors. This package provides the definitions of the predefined types and functions of the language for performing input/output operations on text files. The package header contents shown here are from the 1993 IEEE Standard VHDL Language Reference Manual [8]. The implementation of this package will be consistent across all vendors. The listing of the package header shown next omits the definition of the operators for each type.

```
-------------------------------------------------------------------------------
-- ANSI/IEEE Std 1076–1993*
-- IEEE Standard VHDL Language Reference Manual
```

-- Copyright ©1993 by the Institute of Electrical and Electronics Engineers, Inc.
-- The IEEE disclaims any responsibility or liability resulting from the placement
-- and use in the described manner. Information is reprinted with the permission
-- of the IEEE
--

----
package TEXTIO **is**
-- Type Definitions for Text I/O
--

    type LINE **is access** STRING;   -- A LINE is a pointer to a STRING value.
    type TEXT **is file of** STRING;   -- A file of variable-length ASCII records.
    type SIDE **is** (RIGHT, LEFT);   -- For justifying output data within fields.
    subtype WIDTH **is** NATURAL;   -- For specifying widths of output fields.
--

-- Standard text files:
-- Note these are different from VHDL'87
--

    file INPUT: TEXT **open** READ_MODE **is** "STD_INPUT";
    file OUTPUT: TEXT **open** WRITE_MODE **is** "STD_OUTPUT";
--

-- Input routines for standard types:

    procedure READLINE (**file** F: TEXT; L: **out** LINE);
    procedure READ (L: **inout** LINE; VALUE: **out** BIT;   GOOD: **out** BOOLEAN);
    procedure READ (L: **inout** LINE; VALUE: **out** BIT);
    procedure READ (L: **inout** LINE; VALUE: **out** BIT_VECTOR; GOOD: **out** BOOLEAN);
    procedure READ (L: **inout** LINE; VALUE: **out** BIT_VECTOR);
    procedure READ (L: **inout** LINE; VALUE: **out** CHARACTER; GOOD: **out** BOOLEAN);
    procedure READ (L: **inout** LINE; VALUE: **out** CHARACTER);
    procedure READ (L: **inout** LINE; VALUE: **out** INTEGER; GOOD: **out** BOOLEAN);
    procedure READ (L: **inout** LINE; VALUE: **out** INTEGER);
    procedure READ (L: **inout** LINE; VALUE: **out** REAL; GOOD: **out** BOOLEAN);
    procedure READ (L: **inout** LINE; VALUE: **out** REAL);
    procedure READ (L: **inout** LINE; VALUE: **out** STRING; GOOD: **out** BOOLEAN);
    procedure READ (L: **inout** LINE; VALUE: **out** STRING);
    procedure READ (L: **inout** LINE; VALUE: **out** TIME; GOOD: **out** BOOLEAN);
    procedure READ (L: **inout** LINE; VALUE: **out** TIME);

-- Output routines for standard types

    procedure WRITELINE (**file** F: TEXT; L: **inout** LINE);
    procedure WRITE (L: **inout** LINE; VALUE: **in** BIT;
        JUSTIFIED: **in** SIDE := RIGHT; FIELD: **in** WIDTH := 0);

```
    procedure WRITE (L: inout LINE; VALUE: in BIT_VECTOR;
              JUSTIFIED: in SIDE := RIGHT; FIELD: in WIDTH := 0);
    procedure WRITE (L: inout LINE; VALUE: in BOOLEAN;
              JUSTIFIED: in SIDE := RIGHT; FIELD: in WIDTH := 0);
    procedure WRITE (L: inout LINE; VALUE: in CHARACTER;
              JUSTIFIED: in SIDE := RIGHT; FIELD: in WIDTH := 0);
    procedure WRITE (L: inout LINE; VALUE: in INTEGER;
              JUSTIFIED: in SIDE := RIGHT; FIELD: in WIDTH := 0);
    procedure WRITE (L: inout LINE; VALUE: in REAL;
  JUSTIFIED: in SIDE := RIGHT; FIELD: in WIDTH := 0; DIGITS: in NATURAL:=0);
    procedure WRITE (L: inout LINE; VALUE: in STRING;
              JUSTIFIED: in SIDE := RIGHT; FIELD: in WIDTH := 0);
    procedure WRITE (L: inout LINE; VALUE: in TIME;
          JUSTIFIED: in SIDE := RIGHT; FIELD: in WIDTH := 0; UNIT : in TIME;= ns);
end TEXTIO;
```

B.3   The Standard Logic Package

This package defines the types and supporting functions for the implementation of the IEEE 1164 value system. Most, if not all, vendors make it available and place it in the library IEEE.

```
-- -------------------------------------------------------------------------------------------
-- IEEE Std 1164–1993*
-- IEEE Standard Multivalue Logic System for VHDL Model Interoperability
-- Copyright © 1993 by the Institute of Electrical and Electronics Engineers, Inc.
-- The IEEE disclaims any responsibility or liability resulting from the placement
-- and use in the described manner. Information is reprinted with the permission
-- of the IEEE
--
-- -------------------------------------------------------------------------------------------
-- Title        : std_logic_1164 multi-value logic system
-- Library      : This package shall be compiled into a library
--               : symbolically named IEEE.
--               :
-- Developers   : IEEE model standards group (par 1164)
-- Purpose      : This packages defines a standard for designers
--               : to use in describing the interconnection data types
--               : used in vhdl modeling.
--               :
```

```
--  Limitation     : The logic system defined in this package may
--                  : be insufficient for modeling switched transistors,
--                  : since such a requirement is out of the scope of this
--                  : effort. Furthermore, mathematics, primitives,
--                  : timing standards, etc. are considered orthogonal
--                  : issues as it relates to this package and are therefore
--                  : beyond the scope of this effort.
--                  :
--  Note           : No declarations or definitions shall be included in,
--                  : or excluded from this package. The "package declaration"
--                  : defines the types, subtypes and declarations of
--                  : std_logic_1164. The std_logic_1164 package body shall be
--                  : considered the formal definition of the semantics of
--                  : this package. Tool developers may choose to implement
--                  : the package body in the most efficient manner available
--                  : to them.
--                  :
-- -----------------------------------------------------------------
--  modification history :
-- -----------------------------------------------------------------
-- version | mod. date:|
--  v4.200 | 01/02/91 |
-- -----------------------------------------------------------------


PACKAGE std_logic_1164 IS
    -----------------------------------------------------------------
    -- logic state system (unresolved)
    -----------------------------------------------------------------
    TYPE std_ulogic IS ( 'U', -- Uninitialized
                         'X', -- Forcing Unknown
                         '0', -- Forcing 0
                         '1', -- Forcing 1
                         'Z', -- High Impedance
                         'W', -- Weak    Unknown
                         'L', -- Weak    0
                         'H', -- Weak    1
                         '-'  -- Don't care
                );
    -----------------------------------------------------------------
    -- unconstrained array of std_ulogic for use with the resolution function
    -----------------------------------------------------------------
```

TYPE std_ulogic_vector **IS ARRAY** (NATURAL **RANGE** <>) **OF** std_ulogic;

-- resolution function

FUNCTION resolved (s : std_ulogic_vector) **RETURN** std_ulogic;

-- *** industry standard logic type ***

SUBTYPE std_logic **IS** resolved std_ulogic;

-- unconstrained array of std_logic for use in declaring signal arrays
----
TYPE std_logic_vector **IS ARRAY** (NATURAL **RANGE** <>) **OF** std_logic;

-- common subtypes

SUBTYPE X01   **IS** resolved std_ulogic **RANGE** 'X' **TO** '1'; -- ('X','0','1')
SUBTYPE X01Z   **IS** resolved std_ulogic **RANGE** 'X' **TO** 'Z'; -- ('X','0','1','Z')
SUBTYPE UX01   **IS** resolved std_ulogic **RANGE** 'U' **TO** '1'; -- ('U','X','0','1')
SUBTYPE UX01Z **IS** resolved std_ulogic **RANGE** 'U' **TO** 'Z'; -- ('U','X','0','1','Z')

-- overloaded logical operators

FUNCTION "and" (l : std_ulogic; r : std_ulogic) **RETURN** UX01;
FUNCTION "nand" (l : std_ulogic; r : std_ulogic) **RETURN** UX01;
FUNCTION "or"   (l : std_ulogic; r : std_ulogic) **RETURN** UX01;
FUNCTION "nor" (l : std_ulogic; r : std_ulogic) **RETURN** UX01;
FUNCTION "xor" (l : std_ulogic; r : std_ulogic) **RETURN** UX01;
-- function "xnor" (l : std_ulogic; r : std_ulogic) return ux01;
FUNCTION "not" (l : std_ulogic   ) **RETURN** UX01;

-- vectorized overloaded logical operators

FUNCTION "and" (l, r : std_logic_vector) **RETURN** std_logic_vector;
FUNCTION "and" (l, r : std_ulogic_vector) **RETURN** std_ulogic_vector;
FUNCTION "nand" (l, r : std_logic_vector) **RETURN** std_logic_vector;
FUNCTION "nand" (l, r : std_ulogic_vector) **RETURN** std_ulogic_vector;
FUNCTION "or"   (l, r : std_logic_vector) **RETURN** std_logic_vector;
FUNCTION "or"   (l, r : std_ulogic_vector) **RETURN** std_ulogic_vector;
FUNCTION "nor" (l, r : std_logic_vector) **RETURN** std_logic_vector;
FUNCTION "nor" (l, r : std_ulogic_vector) **RETURN** std_ulogic_vector;

FUNCTION "xor" (l, r : std_logic_vector) **RETURN** std_logic_vector;
FUNCTION "xor" (l, r : std_ulogic_vector) **RETURN** std_ulogic_vector;
-- ---
-- Note : The declaration and implementation of the "xnor" function is
-- specifically commented until at which time the VHDL language has been
-- officially adopted as containing such a function. At such a point,
-- the following comments may be removed along with this notice without
-- further "official" ballotting of this std_logic_1164 package. It is
-- the intent of this effort to provide such a function once it becomes
-- available in the VHDL standard.
-- ---
-- function "xnor" (l, r : std_logic_vector) return std_logic_vector;
-- function "xnor" (l, r : std_ulogic_vector) return std_ulogic_vector;
FUNCTION "not" (l : std_logic_vector) **RETURN** std_logic_vector;
FUNCTION "not" (l : std_ulogic_vector) **RETURN** std_ulogic_vector;
-- ---
-- conversion functions
-- ---
FUNCTION To_bit   (s : std_ulogic;   xmap : BIT := '0') **RETURN** BIT;
FUNCTION To_bitvector (s : std_logic_vector ; xmap : BIT := '0') **RETURN**
BIT_VECTOR;
FUNCTION To_bitvector (s : std_ulogic_vector; xmap : BIT := '0') **RETURN**
BIT_VECTOR;
FUNCTION To_StdULogic   (b : BIT   ) **RETURN** std_ulogic;
FUNCTION To_StdLogicVector (b : BIT_VECTOR) **RETURN** std_logic_vector;
FUNCTION To_StdLogicVector (s : std_ulogic_vector) **RETURN** std_logic_vector;
FUNCTION To_StdULogicVector (b : BIT_VECTOR   ) **RETURN** std_ulogic_vector;
FUNCTION To_StdULogicVector (s : std_logic_vector) **RETURN** std_ulogic_vector;

-- ---
-- strength strippers and type convertors
-- ---
FUNCTION To_X01 (s : std_logic_vector) **RETURN** std_logic_vector;
FUNCTION To_X01 (s : std_ulogic_vector) **RETURN** std_ulogic_vector;
FUNCTION To_X01 (s : std_ulogic   ) **RETURN** X01;
FUNCTION To_X01 (b : BIT_VECTOR   ) **RETURN** std_logic_vector;
FUNCTION To_X01 (b : BIT_VECTOR   ) **RETURN** std_ulogic_vector;
FUNCTION To_X01 (b : BIT   ) **RETURN** X01;
FUNCTION To_X01Z (s : std_logic_vector) **RETURN** std_logic_vector;
FUNCTION To_X01Z (s : std_ulogic_vector) **RETURN** std_ulogic_vector;
FUNCTION To_X01Z (s : std_ulogic   ) **RETURN** X01Z;
FUNCTION To_X01Z (b : BIT_VECTOR   ) **RETURN** std_logic_vector;

```
FUNCTION To_X01Z ( b : BIT_VECTOR     ) RETURN std_ulogic_vector;
FUNCTION To_X01Z ( b : BIT            ) RETURN X01Z;
FUNCTION To_UX01 ( s : std_logic_vector ) RETURN std_logic_vector;
FUNCTION To_UX01 ( s : std_ulogic_vector ) RETURN std_ulogic_vector;
FUNCTION To_UX01 ( s : std_ulogic      ) RETURN UX01;
FUNCTION To_UX01 ( b : BIT_VECTOR      ) RETURN std_logic_vector;
FUNCTION To_UX01 ( b : BIT_VECTOR      ) RETURN std_ulogic_vector;
FUNCTION To_UX01 ( b : BIT             ) RETURN UX01;
-----------------------------------------------------------------
-- edge detection
-----------------------------------------------------------------
FUNCTION rising_edge (SIGNAL s : std_ulogic) RETURN BOOLEAN;
FUNCTION falling_edge (SIGNAL s : std_ulogic) RETURN BOOLEAN;
-----------------------------------------------------------------
-- object contains an unknown
-----------------------------------------------------------------
FUNCTION Is_X ( s : std_ulogic_vector ) RETURN BOOLEAN;
FUNCTION Is_X ( s : std_logic_vector ) RETURN BOOLEAN;
FUNCTION Is_X ( s : std_ulogic       ) RETURN BOOLEAN;
END std_logic_1164;
```

B.4   Other Useful Packages

Users should be aware that vendors will provide other packages that encapsulate many useful functions, such as those for arithmetic and for handling real numbers, as well as miscellaneous utility functions such as type conversion. Some of these packages are vendor specific, while others are currently subject to efforts to arrive at some standards of use.

Check the vendor documentation for other packages that may be available on your system. It is useful to browse through the package headers to obtain an idea of the sets of functions, procedures, or data types that these packages support and thereby understand the motivation for their development. Packages for mathematical functions and type conversion are perhaps the first themes that come to mind. When we think in terms of hardware design, several other needs also become evident. For example, given a specific gate-level design, we may have packages containing various implementation alternatives for the same set of components. One package may contain models for the high-speed implementation of the components, while another may contain models corresponding to the low-power implementation of the same components. The structuring mechanism provided by packages is put to good use in creating libraries of component models used within an organization. Often, these packages are proprietary products of the parent organization.

A Starting Program Template

This appendix serves as a quick reference guide to the structure of a first VHDL model. A template for a general VHDL model can help with the syntactical arrangement of programming constructs. It is useful when trying to remember where to place statements within a program relative to other program constructs. The goal here is to provide a template that contains the most basic and common (and therefore, for our purposes, important) language features—one that will enable the reader to proceed rapidly to the construction of useful VHDL models.

C.1 Construct a Schematic

The first step is the construction of a schematic of the system being modeled.

Construct_Schematic

1. Represent each component (e.g., gate) of the system to be modeled as a *delay element*. The delay element simply captures all of the delays associated with the computation represented by the component and propagation of signals through the component. For each output signal of a component, associate a specific value of delay through the component for that output signal.
2. Draw a schematic interconnecting all of the components. Label each component uniquely.
3. Identify the input signals of the system as input ports.
4. Identify the output signals of the system as output ports.

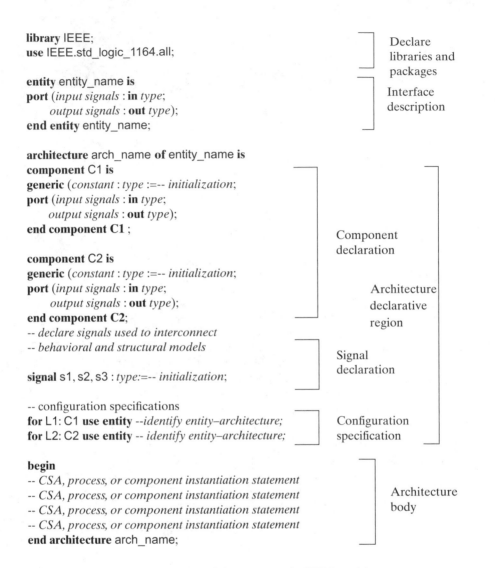

5. All remaining signals are internal signals and should be uniquely labeled.

6. Associate a type, such as **bit**, **bit_vector**, or std_logic_vector, with each input port, output port, and internal signal.

7. Ensure that each input port, output port, and internal signal is labeled with a unique name.

This schematic can now be translated into a VHDL model containing behavioral and structural models of the components make up the system. In fact, the architecture body shown in Figure C.1 can be structured as a series of program statements.

Each statement can be one of the following:

1. A concurrent signal assignment statement

   - simple signal assignment
   - conditional signal assignment
   - selected signal assignment

2. A process

   The process may have a sensitivity list and may consist of a large block of sequential code. Recall that a process execution takes no simulation time and may produce events on signals that are scheduled at some time in the future.

3. A component instantiation statement

   If components have been declared in addition to signals, these components may be instantiated and their input and output ports mapped to signals declared in the architecture. In this manner, these components can be "connected" to, or communicate with, other components, CSAs, or processes.

   This leads to the next procedure for constructing general models reflecting the behavior of the digital system.

C.2   Construct the Behavioral Model

Now we are ready to write the behavioral model.

1. At this point, I recommend using the IEEE 1164 value system. To do so, include the following two lines at the top of your model declaration:

   library IEEE;
   use IEEE.std_logic_1164.all;

   Single-bit signals can be declared to be of type std_logic, while multibit quantities can be declared to be of type std_logic_vector.

2. Select a name for the entity (entity_name) for the system, and write the entity description specifying each input or output signal port, its mode, and associated type.

3. Select a name for the architecture (arch_name) and write the architecture description. Within the architecture description, name and declare all of the internal signals used to connect the components. These signal names are shown on your schematic. The architecture declaration states the type of each signal and, possibly, an initial value.

4. For each delay element, decide whether the behavior of the block will be described by concurrent signal assignment statements, processes, or a component instantiation statement. Depending upon the type, perform the following:

   4.1 *CSA*: For each output signal of the component, select a concurrent signal assignment statement that expresses the value of this signal as a function of the signals that are inputs to that component. Use the value of the propagation delay through the component provided for that output signal. The output signal or one or more input signals (or both) may be a port of the entity.

4.2 *Process*: Alternatively, if the computation of the signal values at the outputs of the component are too complex to represent with concurrent signal assignment statements, describe the behavior of the component with a process. You can use one or more process to compute the values of the output signals from that component. For each process, perform the following:

4.2.1 Label the process. If you are using a sensitivity list, identify the signals that will activate the process.

4.2.2 Declare variables used within the process.

4.2.3 Write the body of the process, computing the values of output signals and the relative time at which these output signals assume those values. If a sensitivity list is not used, insert wait statements at appropriate points in the process to specify when the process should suspend and when it should resume execution. It is an error to have both a sensitivity list and a wait statement within the process.

4.2.4 Complete the process with a set of signal assignment statements, assigning the computed values to the output signals. These output signals may be signals internal to the architecture or may be port signals found in the entity description.

4.3 *Component Instantiation*: For those components for which entity–architecture pairs exist, perform the following:

4.3.1 Construct component declarations for each unique component that will be used in the model. A component declaration can easily be constructed from the component's entity description. For example, the port list will be identical to the port list in the entity description.

4.3.2 Within the declarative region of the architecture description (i.e., before the **begin** statement), list the component declarations.

4.3.3 Within the declarative region of the architecture description (i.e., before the **begin** statement), list the configuration specification if you are not using the default bindings for the component entities.

4.3.4 Write the component instantiation statement. The label is derived from the schematic, followed by the **port map** construct. The port map statement will have as many entries as there are ports on the component. If necessary, include a **generic map** statement.

5. If there are signals that are driven by more than one source, the type of each signal must be a resolved type. This type must have a resolution function declared for use with signals of this type. For our purposes, use the IEEE 1164 types std_logic for single-bit signals and std_logic_vector for bytes, words, or multibit quantities. These are resolved types. Make sure that you include the **library** clause and the **use** clause to capture all of the definitions provided in the std_logic_1164 package.

6. If you are using any functions or type definitions provided by a third party, make sure that you have declared the appropriate library using the **library** clause and declared the use of this package via the presence of a **use** clause in your model.

These steps will produce a fairly generic model. In particular, this approach implies that all design units (entity, architecture, and configuration information) are placed in one physical file. This is clearly not necessary. For example, we know from Chapter 8 that configurations are distinct design units that may be described separately. However, it is often easier to start in the fashion shown here. As our expertise grows, we will be able to avail ourselves of the advantages of dealing with design units separately and managing them effectively. Finally, note that in Figure C.1, the location of the packages is shown as library IEEE. Depending on the packages, that may not be the case, and when writing models, we must have knowledge of the location of any vendor-supplied packages that we are using. We may also be creating our own libraries for retaining user packages.

Index